Experimental Biotechnology

— A Laboratory Manual

NIPA® GENX ELECTRONIC RESOURCES & SOLUTIONS P. LTD.
New Delhi-110 034

Experimental Biotechnology

— A Laboratory Manual

Sunita Dutta
Reader, Department of Zoology
Women's College, Ranchi

Abhijit Dutta
Reader, Department of Zoology

Ashok Kumar Choudhary
Professor, Department of Botany

University of Ranchi (Jharkhand)

NIPA® GENX ELECTRONIC RESOURCES & SOLUTIONS P. LTD.
New Delhi-110 034

NIPA® GENX ELECTRONIC RESOURCES & SOLUTIONS P. LTD.

101,103, Vikas Surya Plaza, CU Block
L.S.C. Market, Pitam Pura, New Delhi-110 034
Ph : +91 11 27341616, 27341717, 27341718
E-mail: newindiapublishingagency@gmail.com
www: www.nipabooks.com

For customer assistance, please contact
Phone: + 91-11-27 34 17 17
Fax: + 91-11- 27 34 16 16
E-Mail: feedbacks@nipabooks.com

ISBN: 978-81-19103-56-0

Composed and Designed by NIPA®.

Preface

Science and technology progressed by leaps and bounds in this century, particularly in the field of Biotechnology and Molecular Biology. Keeping pace with these advances UGC has restructured its curricula and has introduced new experiments at undergraduate and postgraduate levels in various universities and professional courses. Every University and institutions have introduced "Biotechnology" in their syllabus, either as compulsory paper or as vocational paper. It has thus become relevant that infrastructure facilities in life science should be provided in different colleges and universities so that our students should be able to compete nationally and internationally. Yet, for reasons not known there lies a wide range of disparity between the theoretical and practical studies. As a result both teachers and students of undergraduate and post graduate level face tremendous difficulties in acquiring study material of practical utility, particularly topics related to DNA Isolation, Plasmid Isolation, Competent cell preparation, Cloning experiments, PCR, SDS-PAGE, separation techniques for DNA, RNA and proteins.

After the success of our previous book "Experimental Biology-A Laboratory Manual", which presents both the principles of analytical biology that makes experimental, analytical and separation techniques possible in the laboratory, with background information to understand what the students are going to do and why, including technical information the students gather in the laboratory, we thought it would be logical to start our new venture on "**Experimental Biotechnology-A Laboratory Manual**". The book is subdivided into seven sections This encompass: General procedures, like methods of pipetting, solution preparation, buffers and principles of common analytical instruments essential for laboratory biotechnology experiments. The book also includes working with nucleic acid, bacteria, enzymes, proteins; Cloning experiments and a few protocols on Plant biotechnology. Emphasis have been given on DNA/RNA isolation from various sources, use of restriction enzymes, ligation techniques, cloning protocols, screening of transformed cells, various electrophoresis techniques, PCR protocol, etc. The appendices in the last part are included to provide information important to the study of the above mentioned practical as a whole.

This laboratory is the outcome of teaching in this area for past thirty years to M.Sc. and B.Sc. Honours students of life science and those belonging to Biotechnology, Agriculture and allied fields. The idea behind this practical manual was thus to provide theoretical basis of the practical study items to be undertaken in the laboratory in a lucid manner. We shall be grateful if this book is accepted as a valued practical textbook by the students of Life Science at the undergraduate and postgraduate level of most Indian Universities.

❑❑❑

Abbreviation

BCIP	:	5-Bromo-4-Chloro-3'-Indolyphosphate p-Toluidine
BSA	:	Bovine serum albumin
CAP	:	Calf alkaline phosphatase
CsCl	:	Cicium chloride
DMSO	:	Dimethyl sulfoxide
dNTPs	:	Deoxynucleoside triphosphates
dsDNA	:	Double-stranded DNA
EDTA	:	Ethylenediaminetetraacetic acid
EtBr	:	Ethidium Bromide
EtOH	:	Ethidium Hydroxide
GA	:	Gibberellic acid
GTE	:	Glucose Tris EDTA
IAA	:	Indol- 3 acetic acid
LDH	:	Lactate Dehydrogenase
LiCl	:	Lithium Chloride
MOPS	:	3-(N-morpholino)propanesulfonic acid
MS	:	Murashige and Skoog basal salt mixture
NBT	:	Nitro-Blue Tetrazolium Chloride
OD	:	Optical Density
PAGE	:	Polyacrylamide Gel Electrophoresis
PBS	:	Phosphate Buffered Saline

PCIA	:	Phenol-chloroform isoamylalcohol
PCR	:	Polymerase Chain Reaction
PEG	:	Polyethylglycol
SDS	:	Sodium Dodecyl Sulfate
SAP	:	Shrimp alkaline phosphatase
TAE	:	Tris-Acetate-EDTA
Taq	:	*Thermus aquaticus*
TBE	:	Tris-Boric Acid-EDTA
TBS	:	Tris Buffer Saline
TBST	:	Tris-Buffered Saline Tween-20
TE	:	Tris EDTA
TEMED	:	Tetramethylethylenediamine
TSS	:	Transformation and storage solution
TTE	:	Tris/ taurine/ETDA

❐❐❐

Contents

Section-5 : DNA / RNA Separation

Section-6 : Proteins

Section-7 : Plant Biotechnology

□□□

Section-1

General Procedures

Chapter 1

Introduction to the Laboratory

This book is designed to introduce to the students some of the most widely used experimental procedures in biotechnology, including DNA/RNA isolation, manipulation, cloning, protein purification and characterization. The students will also get familiarized with some of the equipments frequently used in biochemistry and molecular biology.

The biotechnology laboratory course is an exploration of procedures. Hence, in order to get full benefit from the course, the students need to read the manual, which is self explanatory. The more effort they put into the course work, the more they will learn. Before the start of actual lab works, one needs to spend some time reading the Laboratory Manual. This reading will provide background information and an outline of the procedures to be performed.

The biotechnology laboratory is conducted as a "directed" research project. This means that although the general procedures are well established, the overall goal of each experiment is the acquisition of new information. Because of the nature of scientific research, predicting the outcome of experiments that have not previously been performed is difficult. It may therefore be necessary to design new experiments based on the results of previous ones, or to repeat experiments that yielded uninterruptable or ambiguous results.

Laboratory Safety

Laboratories contain hazards of various kinds. Students are required to wear shoes, trousers and a lab coat inside the laboratory. Without instructor students should not work in the laboratory. Some of the chemicals used are toxic, mutagenic, or teratogenic; hence students should take maximum precautions while handling these hazardous chemicals.

Keeping a Laboratory Notebook

Students are required to maintain a laboratory notebook for the recording of laboratory data and calculations, and are important for writing the lab reports. In the notebook, each experiment should begin with a title, a date, and a statement of the objective of the planned work. One should also record exactly what one did at each step. In addition, one needs to record various numerical informations like weights of reagents, absorbance readings, enzyme activities, protein concentrations, and buffer concentrations, etc.

Molecular Biology

Molecular biology in its modern meaning (*i.e.* referring to genetic manipulation and analysis techniques) is a fairly new science; nearly all of the techniques used were invented after 1970. This means that techniques are still being invented, and many of the procedures have changed (often dramatically) during the last few years.

In molecular biology, exact measurement of the reagents is of immense importance, like buffer concentration, pH etc.; in others, such as ligation reactions, the DNA concentration can vary within a factor of ten (or more) and still allow the procedure to be successful. Also sometimes the entire experiment of molecular biology requires utmost care.

Purification of proteins from normal cells has its advantages and disadvantages and has been a practice since long. The proteins suitable for study are limited to those expressed by readily available sources, and to those expressed in significant quantities and sometime difficult due to lack of availability of source material. Also experimental modifications can be introduced into the protein but with greater difficulties.

As a result, expression of genes in heterologous organisms, such as bacteria (*Escherichia coli*), is frequently used. Bacterial protein expression generally allows the protein to be expressed in very large quantities, allows the researcher to chose the form of the protein to be expressed, and allows the researcher to introduce mutations in the protein sequence to examine the roles of individual amino acids in the function of the protein.

E.coli, found in the intestinal tracts of a many species of mammals, including man, has been used for a wide variety of molecular biological experiments. Techniques for manipulating DNA in *E. coli* are well established.

We (and many other organisms) secrete enzymes that contaminate nucleic acids. DNase (the enzyme that degrades DNA) is rapidly denatured by heating at 68°C; for this reason, it is good ideas to heat treat the DNA samples if there is any chance of DNase contamination.

Many molecular biological procedures are very commonly used and all of the required reagents are available in kit form. These kits have made many aspects of molecular biology much easier.

Chapter 2

Pipetting

In molecular biology and biochemistry, the ability to accurately and reproducibly measure and transfer small volumes of liquids is critical for obtaining useful results. For volumes less than 1 ml, the most common method for measuring liquid volumes involves the use of a device known as a pipetman or micro-pipette.

The pipetmen can be both fixed and variable volumes. Commonly used pipetmen are: P1000, P200, P20, P10 etc. P1000 are useful for volumes from 200 to 1000 µl. P200 are useful for volumes from 20 to 200 µl. P20 are useful for volumes from 0.5 to 20 µl. For beginners fixed volume pipetmen are more useful. It is important to ascertain whether the correct pipetman is chosen for measuring the volume one needs.

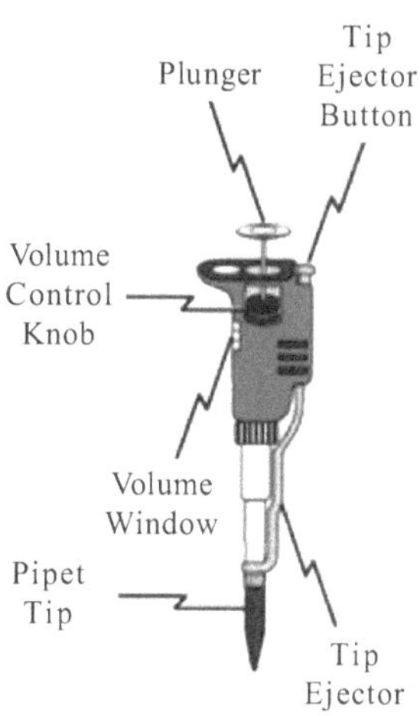

Fig.1: Parts of a Micro-Pipette

Disposable tips are used for all pipetmen (pipettes without appropriate tips should be avoided while pipetting liquids, because it might contaminate the pipetman and may damage it). When attaching the tip, make certain that the tip is the correct type for the pipetman and that the tip is properly seated on the end of the pipetman.

While depressing the plunger depresses, one will feel a sudden increase in resistance. This is the first "stop". If depressing is continued, one will find a point where the plunger no longer moves downward, "the second "stop". While using the pipet, the plunger should be depressed to the first "stop", placing the tip into the liquid, and in a slow, controlled manner, allow the plunger to move upwards.

Next, the pipetman, with the pipetted liquid in the tip, should be inserted into the container to which one wishes to add liquid. The plunger should be depressed to the first, and then to the second stop. Depressing to the second stop expels all of the liquid from the tip.

Testing the Accuracy of Pipetting

It is important to practice pipetting to develop good technique. To do this students are advised to pipet water and weigh the volume to determine accuracy. Since the density of water is 1.0 g / ml, the mass of the volume of water pipetted in grams is equal to its volume in ml. This simple conversion will be used to determine the accuracy of pipetting.

Procedure

1) Take a piece of weigh paper on the balance and tare it so that the scale reads 0.000 g.
2) Set the pipetman -1000 for 1000 µl and put a tip on it. Push down the plunger, put it in the water and pull up slowly. Put this amount of water on the scale and weight it. Record this weight on a piece of paper.
3) Discard the water from the weight paper and tare it again to 0.000 g.
4) Use the same pipetman, rinse the tip with water. Do this ten times slowly. Now push down the plunger and allow it to jump up quickly. Weigh this amount of water and record it.
5) Discard the water and re-tare the balance. This time pull the water into the pipette tip very slowly. Weigh this amount of

water. If this doesn't weigh 1.000 (+/- 0.005) g, then repeat this slowly until you get this degree of accuracy.

6) Repeat step five with two other volumes. Good volumes to use are 0.500 ml and 0.200 ml.

7) Do step five using the P-200 at 200 μl, 100 μl, and 25 μl, until you get accurate results (< 0.5% error).

8) Do step five using the P-20 at 20.0 μl, 10.0 μl, and 5.0 μl, until you get accurate results.

Chapter 3

Measuring Absorbance

A spectrophotometer is an analytical instrument that measures the absorbance of a test sample (solution). The term "spectroscopy" comes from the word "spectrum". "Spectroscopy" therefore implies the use of multiple wavelengths of light. Spectrophotometers have the ability to specifically measure absorbance at specific wavelengths. The most commonly used method to allow this involves a "monochromator", a device (either a prism, or more commonly, a diffraction grating) that splits the incident light into its component wavelengths, and allows only light of the desired wavelength to reach the sample. The ability to measure absorbance at different wavelengths is very useful, because the extinction coefficient of a compound varies with wavelength. In addition, the absorbance spectrum of a compound can vary depending on the chemical composition of the compound, and depending on the nature of the solvent around the compound.

Absorbance is a useful quantity and follows the Beer-Lambert law. The extinction coefficient of a molecule at a given wavelength can be calculated using the Beer-Lambert equation from absorbance measurements for solutions of known concentration, which states that:

$A = \Sigma cl$

Where:

A is the absorbance of the sample at a particular wavelength, and is defined as the extinction coefficient for the compound/element at that wavelength measured in M x cm^{-1},

c is the molar Σ concentration of the absorbing species, and

l is the path length of the solution in cm.

If the extinction coefficient of an absorbing species is known, the absorbance of the solution can be used to calculate the concentration of the absorbing species in solution. (This means that the species of interest is the only material that absorbs at the wavelength being measured). Thus, absorbance allows us to calculate the concentration of compounds in solution, although it does not clearly define absorbance. Absorbance can be defined as:

$$A = \log \frac{(I_0)}{(I)}$$

Where:

- I_0 is the amount of light entering the sample, and
- I is the amount of light leaving the sample.

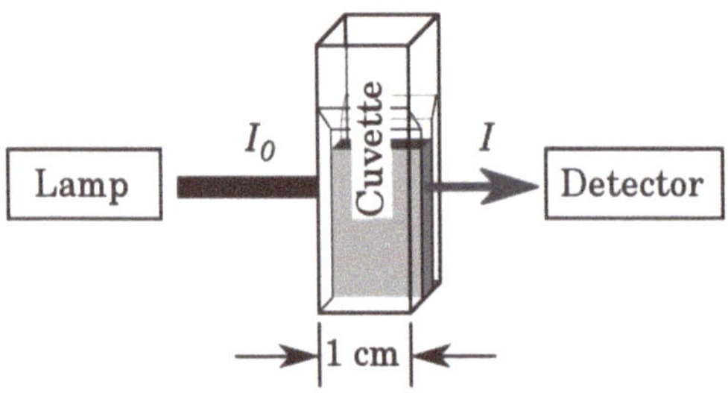

Fig. 2: The typical internal arrangement of a Spectrophotometer

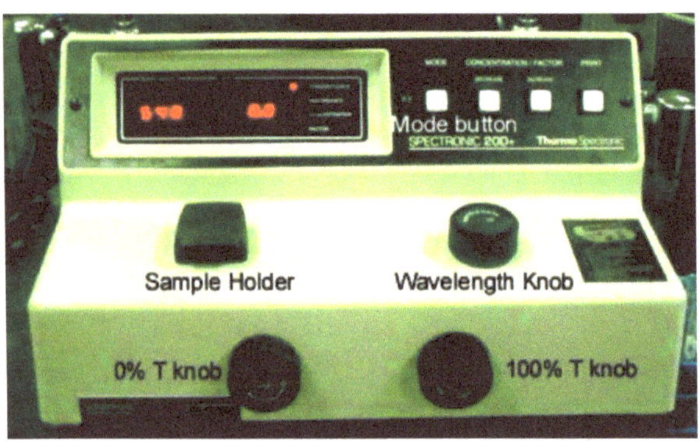

Fig.3: A common Laboratory Spectrophotometer

Absorbance is therefore a measure of the portion of the light leaving the lamp that enters the detector. When absorbance = 1, only 10% of the light reaches the detector; and when absorbance = 2, only 1% of the light makes it to the detector. Absorbance values greater than 2 are defective and unreliable, as very little light will reach the detector to allow accurate measurements. The values should be noted while measuring the absorption. Repeated measurements should be made when the absorbance is greater than 2 to negate erroneous results.

Spectrophotometers measure the decrease in the amount of light reaching the detector. Hence, care should be taken while inserting the cuvettes that there should be no smudges or fingerprints or water marks or air bubbles stuck to the cuvettes. Cuvettes are either square or circular of 1cm dimension.

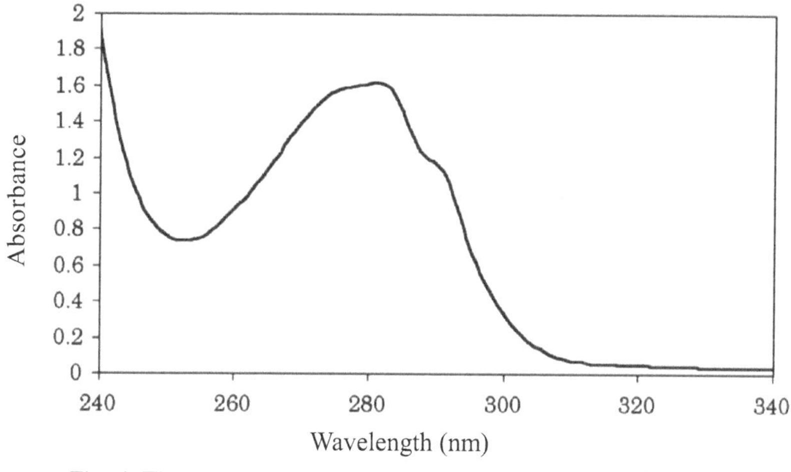

Fig. 4: The absorbance against wavelength graph of protein

The figure above shows the absorbance graph of a protein, having a strong absorbance peak near 280 nm, which diminishes at longer wavelengths. In protein the aromatic amino acids tryptophan and tyrosine are the only chromophores (chemical groups within a compound that absorbs light). Colored proteins, like hemoglobin, exhibit their color due to the presence of heme (chromophores) that absorb in the visible portion of the spectrum.

Chapter 4

Preparing Dilutions

To prepare a series of dilutions of the original more concentrated stock solution, one need to consider both the required final concentration and required volume of the diluted material. The dilution can be readily calculated by the equation:

$$C_1V_1 = C_2V_2$$

Where:

C_1 is the concentration of the initial solution;

V_1 is the volume of the initial solution to be used for dilution

C_2 is the desired final concentration, and

V_2 is the desired final volume.

The initial concentration and the final concentration are either known or are chosen. The final volume is usually an amount that is chosen based on the amount required for a given experiment. This means that at least three of the required terms are either known or can be chosen by the experimenter.

For example, if a stock solution of 1000 μg/ml BSA is taken, and one needs to prepare a diluted solution of 200 μl of 20 μg/ml, therefore, C_1 = 1000 μg/ml; C_2 = 20 μg/ml, and V_2 = 200 μl. This leaves V_1 as the unknown value (*i.e.* how much of the stock solution must be diluted to 200 μl final volume to yield the desired concentration). Rearranging the dilution equation gives:

$$V_1 = V_2 \frac{C_2}{C_1}, \text{ and}$$

$$\text{Therefore, } 4\ \mu l = 200\ \mu l \ \frac{(20\ \mu l/ml)}{(1000\ \mu l/ml)}$$

Thus, one needs to dilute 4 μl of the stock solution to a final volume of 200 μl (i.e. by adding 196 μl).

In case one wishes to make a solution of 1 μg/ml, the same equation would indicate that one need 0.2 μl of the 1000 μg/ml stock solution for 200 μl of the final diluted sample.

Concentration ratios are of considerable value. For example, a stock solution of buffer contains:

- 450 mM Tris-HCl,
- 10 mM EDTA, and
- 500 mM NaCl.

Let us suppose one wants to make a final concentration of 45 mM Tris-HCl, 1 mM EDTA, and 50 mM NaCl. Thus in the concentration of the final buffer is one-tenth that of the original each case. Hence preparing a 1:10 dilution of the stock solution gives the appropriate final concentration of each solution. Therefore, the stock solution of buffer will be called a 10x stock, because it is ten-times more concentrated than the final, useful buffer.

Making Dilutions

Many of you appear to panic when you must dilute something, yet the mathematics involves nothing worse than the simplest algebra. One reason is simply that when you are busy with a laboratory procedure you are distracted and it is difficult to think in the abstract. That problem can be overcome by practicing in advance of the need. Even with practice, though, you may find dilution problems confusing unless you very clearly define your objectives. We will give you a useful formula for making dilutions, one that you may have seen before. The formula is worse than useless, though, if you don't use it properly.

Establish a frame of reference

For the sake of simplicity, let's say we are talking about sucrose solutions. Suppose you have a starting solution of sucrose (in water) with volume V1 and concentration C1. What is the total amount of sucrose in your solution? Answer: C1xV1.

Example 1

Volume = 0.2 liter; concentration is 50 grams/liter. C1xV1 = 50 grams/liter x 0.2 liter= 10 grams sucrose.

Now suppose that you dilute that solution with water—the whole thing—to some larger, predetermined volume (V2). What amount of sucrose is present in the new, diluted solution? If you said 10 grams, you get the gold star. But wait a minute—C1xV1 = 10 grams, and the new solution has a different volume, V2. The same amount of sucrose is present in the new solution as was in the original solution, so the following relationship must hold:

C1xV1 = C2xV2, where C2 = concentration of the new solution.

Example 2

Dilute the previous sucrose solution to 2 liters. What is the concentration of the new solution? We must solve for C2, of course. C1xV1 = C2xV2 = 10 grams. We know that V2 = 2 liters, so now we have

C2 x (2 liters) = 10 grams

Solve for C2 to obtain 5 grams/liter.

Determining what you already knows and putting the information into the equation C1xV1 = C2 x V2 establishes the relationship that you need in order to solve dilution problems.

Determine the objective

What do you want to do? Or, more realistically, what does the instructor want you to do? Two types of dilution problems are quite common in biology and biochemistry labs.

- ❐ Dilute a known volume of known concentration to a desired final concentration
- ❐ Dilute a known concentration to a desired final concentration AND volume

The second type of problem really throws people off! Let's start with the first one, though. You know V1, C1, and C2 is predetermined. It remains, then, to solve for V2, namely the final volume to which to dilute the solution. This one is easy, since you keep the amount of solute the same and only have to change one factor.

It is known that C1 is the concentration of the starting solution. You have predetermined both V2 and C2, namely the final volume and concentration that you desire. There is one undetermined variable left, namely V1. V1 is the volume of original solution that you will dilute to the desired final volume and concentration.

$$C1 \times V1 = C2 \times V2$$

$$V1 = (C2 \times V2)/C1 \text{ or,}$$

$$V1 = (\text{final amount of solute})/C1$$

Example 3

You have a sucrose solution of 47 grams/liter. You want to prepare 100 milliliters (0.1 liter) of sucrose solution of concentration 25 grams/liter. Since you know the starting concentration of sucrose and you know both the final concentration and volume of solution that you want, all you need to find out is what volume of starting solution (V1) to use.

V1 = (C2 x V2)/C1, that is, V1 = (25 grams/liter x 0.1 liter) ÷ 47 grams/liter = 0.053 liter = 53 ml

Notice that the calculation comes out to 0.05319149L using full precision, but it has been rounded off the required volume to the nearest ml. There is a limit to the precision with which we can prepare a solution, and also a limit to the precision that we really need. You would use a 100 ml graduated cylinder to determine final volume. You would be able to read the markings to the nearest 1 ml, as a rule.

Units

It is critical that you report units for concentrations, volumes, and amounts, and when you make calculations for dilutions you must not mix up the units. For example, it doesn't work to write,

V1 (160 milliliters) x C1 (160 milligrams/liter = V2 (unknown) x C2 (desired-3grams/liter).

However, because concentration represents a proportional relationship, you can select from a variety of units. For example, 1 milligram/milliliter is the same as 1 gram/liter, 1 microgram/micro liter, or 1 monogram/nanoliter. It is also the same as 1000 milligrams/liter, but why would we write it that way?

Select units that simplify your expressions

Working with Stock Solutions

We define a "stock solution" as a concentrate, that is, a solution to be diluted to some lower concentration for actual use. We may use just the stock solution or use it as a component in a more complex solution. We refer to the solution that we end up using as a "working solution". If you are comfortable making dilutions then you can appreciate the many advantages of working with stock solutions. Stock solutions can save a lot of time, conserve materials, reduce needed storage space, and improve the accuracy with which we prepare solutions and reagents. Here are several illustrated types of applications using stock solutions.

What do the ratios mean?

Suppose you have been asked to "prepare a one to ten dilution of solution X." Does it mean take one part solution X and add ten parts water, or does it mean take one part solution X and bring the volume to a total of ten parts? We often prepare complex media as weight-to-volume (w: v) instead of weight-in-volume (w/v), that is, we add a prescribed mass of material to a prescribed volume of water instead of mixing the materials and bringing the mixture to a prescribed final volume.

To avoid confusion you might say "please prepare a one to ten dilution of solution X, weight-in-volume," or if you want to bring materials together in a precise proportion, say "please prepare a one to ten dilution of solution X, weight-to-volume." In the latter case it might be less confusing just to say "please combine one part solution X with ten parts water."

Calculations

The concentration units for chemical standard solutions used for ICP applications are typically expressed in µg/mL (micrograms per milliliter) or ng/mL (nanograms per milliliter). For example, a 1000 µg/mL solution of Ca^{+2} contains 1000 micrograms of Ca^{+2} per each mL of solution and a 1 µg/mL solution of Ca^{+2} contains 1000 ng of Ca^{+2} per milliliter of solution. To convert between metric concentration units the following conversions apply:

Table 1: Mass portion of concentration unit where g = gram

Prefix	Scientific Notation	Decimal equivalents	Example Units
kilo- (k)	$= 10^{3}$	= 1000 g	kilogram (kg)
milli- (m)	$= 10^{-3}$	= 0.001 g	milligram (mg)
micro- (µ)	$= 10^{-6}$	= 0.000001 g	microgram (µg)
nano- (n)	$= 10^{-9}$	= 0.000000001 g	nanogram (ng)
pico- (p)	$= 10^{-12}$	= 0.000000000001 g	picogram (pg)

Table 2: Volume portion of concentration unit where L = liter

Prefix	Scientific Notation	Decimal equivalents	Example Units
milli- (m)	$= 10^{-3}$	= 0.001 L	milliliter (mL)
micro- (µ)	$= 10^{-6}$	= 0.000001 L	microliter (µL)
nano- (n)	$= 10^{-9}$	= 0.000000001 L	nanoliter (nL)
pico- (p)	$= 10^{-12}$	= 0.000000000001 L	picoliter (pL)

The difference between ppm and µg/mL is often confused. A common mistake is to refer to the concentration units in ppm as a short cut (parts per million) when we really mean µg/mL. One ppm is in reality equal to 1 µg/g. In similar fashion ppb (parts per billion) is often equated with ng/mL. One ppb is in reality equal to 1 ng/g.

Parts Per Million (PPM)

A unit of concentration often used when measuring levels of pollutants in air, water, body fluids, etc. One ppm is 1 part in 1,000,000. The common unit **g of solute/10^6 g of solution** is equal to ppm. i.e., 1g of solute in 10^6 g of solution is 1ppm.

Parts Per Billion (PPB)

One part per billion is 1 part in 1,000,000,000. i.e., 1g of solute in 10^9 g of solution is 1ppm.

Parts Per Trillion (PPT)

A unit of concentration used to measure vanishingly small levels of pollutants or contaminants in, for example, body fluids. One ppt is 1 part in 1,000,000,000,000. . i.e., 1g of solute in 10^{12} g of solution is 1ppm.

The remarkable advances in the sensitivity of modern analytical techniques make it possible to detect some substances at the ppt level whose presence would not have been detected using earlier assay methods.

To convert between ppm and ppb to μg/mL or ng/mL the density of the solution must be known. The equation for conversion between wt. /wt. and wt. /vol. units is:

(μg/g) (Density in g/mL) = μg/mL

and/or (ng/g) (density in g/mL) = ng/mL

Therefore, if we have a solution that is 1000 μg/mL Ca^{+2} and know or measure the density to be 1.033 g/mL then the ppm Ca^{+2} = (1000 μg/mL) / (1.033 g/mL) = 968 μg/g = 968 ppm.

When making dilutions the following equation is useful:

(mLA)(CA) = (mLB) (CB)

For example, to determine how much of a 1000 μg/mL solution of Ca^{+2} required to prepare 250 mL of a 0.3 μg/mL solution of Ca^{+2} we would use the above equation as follows:

(mLA)(1000 μg/mL) = (250 mL) (0.3 μg/mL)

(mLA) = [(250 mL) (0.3 μg/mL)]/ (1000 μg/mL)

(mLA) = 0.075 mL = 75 μL

Molarities and Dilutions

One mole is: "The amount of a substance which contains as many elementary entities as there are atoms in 0.012 kilogram of carbon 12".

Since a molecular weight in grams (mole) of any substance contains the same number of molecules, then according to Avogadro's principle, the molar volumes of all gases should be the same at same temperature and pressure. The number of molecules in one mole is now called Avogadro's number. Avogadro had no knowledge of moles, or of the number that was to bear his name, which was never actually determined by Avogadro himself.

Avogadro's number is: 602,000,000,000,000,000,000,000 (6.02×10^{23})

How big is Avogadro's number?

- An Avogadro's number of soft drink cans would cover the surface of the earth to a depth of over 200 miles.
- If we were able to count atoms at the rate of 10 million per second, it would take about 2 billion years to count the atoms in one mole.

A solution is a homogeneous mixture where all particles exist as individual molecules or ions. The molarity of a solution is calculated by taking the moles of solute and dividing by the litres of solution:

$$\text{Molarity} = \frac{\text{Moles of solute}}{\text{Litres of solution}}$$

The molality (m) of a solution is the concentration measured as moles of solute per kilogram of solvent.

$$\text{Molality} = \frac{\text{No. of Moles of solute}}{\text{Wt. of solvent in Kg}}$$

For example, a 1 m (not a 1 M) NaCl solution contains 1 mole of NaCl per kilogram of water. Molalities are preferred over molarities in experiments that involve temperature changes of solutions, e. g. calorimetry and freezing point depression experiments. Molarity = moles per litre, molality = moles per kilogram.

Example 1

What is the molarity of 2 moles of solute dissolved in 1 litre of solvent?

$$\text{Molarity} = \frac{\text{2 moles}}{\text{1 litre}} = \text{2 moles / L} = \text{2M}$$

i.e. 2 moles/litre, or "2 molar".

Example 2

What is the molarity of 0.75 moles of solute dissolved in 2.5 litres of solvent?

$$\text{Molarity} = \frac{0.75 \text{ mol}}{2.5\text{L}} = 0.3\text{M}$$

Example 3

What is the molarity of 40 grams of NaOH dissolved in 2 litres of solvent? This calculation must be performed in two stages:

a. Convert grams to moles
b. Divide moles by litres to calculate molarity

The molecular weight of NaOH is 40 grams/mol

$$\frac{40\text{g}}{40\text{g/mol}} = 1\text{mol}$$

$$\frac{1\text{mol}}{2\text{L}} = 0.5\text{M}$$

Equations to be remembered

1. $$\frac{\text{Grams}}{\text{Molecular weight}} = \text{moles (mol)}$$

2. $$\frac{\text{Moles}}{\text{Volume of solutions in litres}} = \text{molarity (M)}$$

3. $$\text{Molarity (M)} = \frac{\text{Grams}}{\text{Molecular weight}}$$

Example 4

When 2 grams of NaCl (molecular weight 58.44 g mol^{-1}) is dissolved in 100 mL of solute, what is the molarity of the solution? (MV=g/ mol.wt.)

$$(x\ M)\ (0.1L) = \frac{2g}{58.44}$$

Therefore:

$$(x\ M)\ (0.1) = 0.034 = 0.34M$$

Example 5

How many grams of NaCl are needed to make 500 ml of a 0.2 M solution? (MV=g/mol.wt.)

$$(0.2\ M)\ (0.5L) = \frac{x}{58.44}$$

Therefore:

$$0.1 = \frac{x}{58.44} = 5.844 \text{ grams}$$

Dilution Calculations

Solutions are prepared sometimes by diluting a more concentrated solution; hence one has to be familiar with the calculations that are associated with dilutions.

It is quite simple to calculate such types of equations and the element of simplicity is that the number of moles of solute stays the same. The number of moles of solute in the concentrated solution (indicated by the subscripted molescon) is equal to the number of moles in the dilute solution. You have simply increased the amount of solvent in the solution.

The number of moles of solute in the concentrated solution is equal to the molarity of the concentrated solution times the volume of the concentrated solution. Also, the number of moles of solute in the dilute solution is equal to the molarity of the dilute solution times the volume of the dilute solution. Since we are really interested in the molarities and volumes we can substitute and use the equation shown in the second line (or third line, depending on your preference for how to show multiplication).

$$\text{Times dilution} = \frac{\text{Final volume of solution}}{\text{Initial volume of solution}}$$

$$\text{Times dilution} = \frac{\text{Initial molarity of solution}}{\text{Final molarity of solution}}$$

Volume of solvent added = Final volume of solution Initial volume of solution

Let's use this equation in a few examples.

$moles_{con} = moles_{dil}$

$moles_{con} \times Vol_{con} = M_{dil} \times Vol_{dil}$

$(M_{con})\ (V_{con}) = (M_{dil})\ (V_{dil})$

Dilution Numericals

1. How much 2.0 M NaCl solution would you need to make 250 mL of 0.15 M NaCl solution?

   Ans. 19 ml (2 s.d. from 18. 75ml)

2. How much 2.0 M NaCl solution would you need to make 250 mL of 0.15 M NaCl solution?

   Ans. 0.76 M (2 s.d.)

3. What would be the concentration of a solution made by adding 250 mL of water to 45.0 mL of 4.2 M KOH?

   Ans. 0.64 M

4. How much 0.20 M glucose solution can be made from 50 mL of 0.50 M glucose solution?

   Ans. 130 mL (2 s.d. from 125 mL)

Answers

Here are the answers to the questions in exercise

$$(M_{con})\ (V_{con}) = (M_{dil})\ (V_{dil})$$

$$(V_{con}) = \frac{(M_{dil})\ (V_{dil})}{(M_{con})}$$

$$(V_{con}) = \frac{(0.050\ M)(500.mL)}{(6.0M)}$$

$$V_{con} = 4.2\ mL$$

Chemist starts with 50.0 mL of a 0.40 M NaCl solution and dilutes it to 1000. mL. What is the concentration of NaCl in the new solution?

$$(M_{con})\,(V_{con}) = (M_{dil})\,(V_{dil})$$
$$(0.40\ M)\,(50.0\ mL) = (M_{dil})\,(1000.\ mL)$$
$$\frac{(0.40\ M)\,(50.0\ mL)}{(1000.\ mL)} = (M_{dil})$$
$$0.020\ M = M_{dil}$$

Summary

- The concentration of a solution is usually given in moles per litre (mol L-1 OR mol/L).

This is also known as molarity.

- Concentration, or Molarity, is given the symbol M.

A short way to write that the concentration of a solution of hydrochloric acid is 0.01 mol/L is to write [HCl] =0.01M

The square brackets around the substance indicate concentration.

- M = n ÷ V (M=concentration of solution in mol/L, n=moles of substance, V=volume of solution in litres (L))

This formula can be re-arranged

- n = M x V (M=concentration of solution in mol/L, n=moles of substance, V=volume of solution in litres (L))
- V = n ÷ M (M=concentration of solution in mol/L, n=moles of substance, V=volume of solution in litres (L))

Examples

1. M = n ÷ V

Calculate the concentration (molarity) of a sodium chloride solution containing 0.125 moles sodium chloride in 0.5 litres of water.

- M = n ÷ V in litres
- n = 0.125mol
- V in litres = 0.5L
- [NaCl(aq)] = M = 0.125 ÷ 0.5 = 0.25M (or 0.25mol/L or 0.25mol L^{-1})

2. *n = M x V*

Calculate the moles of copper sulfate in 250mL of 0.02M copper sulfate solution.

- n = M x V
- M = 0.02M
- V = 250mL = 250 ÷ 1000 = 250 x 10-3L = 0.250L (since there are 1000mL in 1L)
- n = 0.02 x 250 x 10-3 = 0.005mol

3. *V = n ÷ M*

Calculate the volume of a 0.80M potassium bromide solution containing 1.6 moles of potassium bromide.

- V = n ÷ M
- n = 1.6mol
- M = 0.80M
- V = 1.6 ÷ 0.80 = 2.00L

Practice Exercises

1. Make 25 ml of 20% ethanol from a 95% ethanol stock.
2. Make 500 ml of 1X SSC from a 20X SSC stock. (Remember, it doesn't matter what the units are or what the chemicals are. The calculation is the same.)
3. Make 20 μl of 1X restriction enzyme buffer from a 10X stock.
4. Make 20 ml of 0.5% NaCl from a 10% stock.
5. Make 100 ml of a solution that is 5 mM $MgCl_2$, 5 mM $CaCl_2$ and 125 mM KCl from 1 M stocks of each of these chemicals.

Chapter 5

Laboratory Buffers

Several substrates, like proteins (enzymes), are very sensitive to alteration in the concentration of the solution they are placed in. Hence is the importance of a buffer which controls the chemical properties of the medium. Literally a "buffer" has the ability to resist changes in the hydrogen ion concentration. Although, buffers do contain other molecules, that might, in addition to the hydrogen ion concentration, influence the ionic strength, the activity of enzymes, and other parameters of the experiment.

In any biological experiment, appropriate buffer must be selected based on their effect on the experiment. Ideal buffers control pH and ionic strength without interacting adversely with the experiment undertaken. For example, although phosphate is a common buffer, it may not be suited when phosphate is a substrate or product of the reaction being studied. Also, some proteins interact poorly with some buffers. For example, Tris is not ideal as it exhibit high pK*a* and the large change in pK*a* when temperature changes, although it is cheap and most proteins are stable in Tris buffers, moreover it seldom reacts with reaction substrates hence, Tris buffer is commonly used in biochemical experiments.

Honderson-Hasselbach equation

This equation is useful for calculating the pH of a solution and also for predicting whether a particular compound will be useful as a buffer over a given pH range, although, many buffers deviate significantly from ideal Honderson-Hasselbach behavior.

$$pH = pKa \log \frac{[A\hat{I}]}{[HA]}$$ Honderson-Hasselbach equation

Buffers are, thus, prepared to overcome commonly observed deviations from ideal behavior by adjusting the pH solution by adding an acid or a base, and water to achieve the desired volume.

For example,

Prepare 1 liter of 50 mM Tris-HCl buffer (pH 7.4) with 200 mM sodium chloride:

- In a flask add 50 mmoles of Tris base
- 200 mmoles of sodium chloride to a flask and
- Add 900 ml water.
- Add HCl to reduce the pH to 7.4
- Monitor the change in pH by a pH meter, while adding water to yield a 1 liter final volume.

Chapter 6

Operating Systems of Common Instruments

Autoclave Operating Procedures

Place all material to be autoclaved in an autoclavable tray. All items should have indicator tape. Separate liquids and solids should be autoclaved separately. Ensure that the lids on all bottles are loose. Do not put a large number of items in tray, to ensure that all items reach the appropriate temperature. Also allow sufficient air/steam circulation.

Steps in autoclave

1. Pressure should read 'O' when you open the pressure chamber of the autoclave.
2. Place items to be autoclaved in the autoclave and close the door. Some autoclaves require that you also lock the door after it's closed.
3. Set time - typically 20 minutes.
4. Temperature should be set at 121°C already, but double-check and change if necessary.

5. *Set cycle*: If liquid, set "liquid cycle" or "slow exhaust". If dry, set "dry cycle" or "fast exhaust" + dry time.
6. Start the cycle. On some autoclaves, the cycle starts automatically at step 5. On others, turn to "sterilize".
7. At the end of the cycle, check that: a. the chamber pressure is at 0; b. the temp is <100 °C
8. Open door.
9. Remove contents using gloves and immediately tighten all caps.

Operating Instructions for Spectrophotometer

- To measure the absorbance of a solution in the short-wave range (<300 nM) use the quartz cuvettes. Disposable plastic cuvettes are available for reading in the visible range.
- Turn the spectrophotometer on - the switch is on the right in the back.

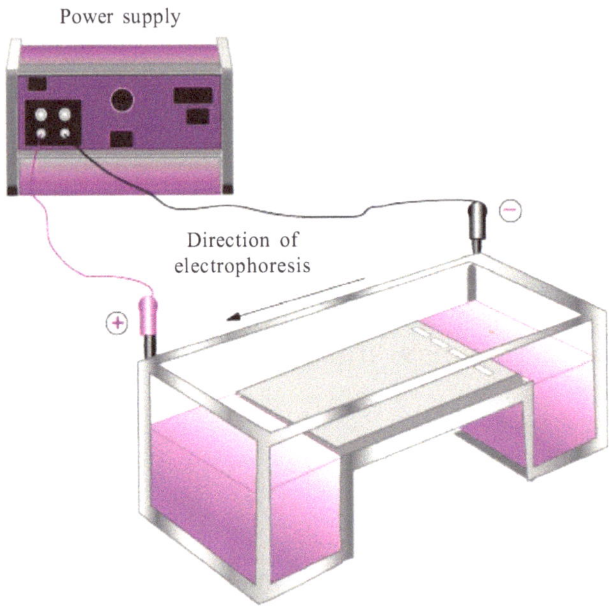

Fig. 5: A horizontal Gel Electrophoresis System for Agarose Electrophoresis showing gel casting, loading of sample and power supply.

- Allow the instrument to calibrate. Do not open the chamber during this time. The deuterium lamp is OFF by default. To read absorbance in the UV range, turn the lamp on as follows after the machine has completed its calibration: Depress the function key. Press the mode key. Press enter. For best accuracy, the lamp should be warmed up for 20 minutes.

- ❒ Press the function key "O" is displayed. Press enter. Using the up or down arrow keys, enter in the desired wavelength.
- ❒ Prepare a reference cuvette containing the same diluents as your sample. Prepare your sample.
- ❒ Place the reference cuvette in cell 1 and place your samples in cells 2-6.
- ❒ Press the cell key until cell 1 is in position. Press the Set Reference key to blank against the appropriate buffer. Press the cell key to advance to read the next sample.

Section-2

Nucleic Acids and *E.coli*

Chapter 7

Working with Nucleic Acids

Storage

Nucleic acids are very sensitive to contaminations hence care should be taken while processing and storing DNA and RNA. The following properties of reagents and conditions need to be taken care of:

- Heavy metals cause breakage of phosphodiester linkage. One can use EDTA which is an excellent heavy metal chelator.
- UV rays cause extensive damage to nucleic acids. UV light at 260 nm causes breakage of thymine dimers and cross-link, including loss of biological activities.
- Ethidium bromide results in photo oxidation of DNA. Oxidation products can cause phosphodiester breakage.
- Nucleases are found on human skin; therefore, it is advisable to avoid direct or indirect contact between nucleic acids and fingers.
- Most DNAes are not very stable; however, many RNAses are very stable and can absorb to glass or plastic and remain active.

- At -20°C: extensive single and double strand breaks results. Hence for long-term storage it should be stores at -70°C.
- It is best to store DNA in high salt (>1M) in the presence of high EDTA

  (>10mM) at pH 8.5, for long-term storage.
- Storage of DNA in CsCl with ethidium bromide in the dark is an excellent storage medium. There is about one phosphodiester break per 200 kb of DNA per year.
- Storage of ë DNA in the phage is better than storing the pure DNA.

Quantitation

1. Spectrophotometric determination

The simplest method of finding the concentration DNA in a pure solutions is by measuring the absorbance at 260 nm where an OD of 1 in a 1 cm path length = 50 μg/ml for double-stranded DNA, 40 μg/ml for single-stranded DNA and RNA and 20-33 μg/ml for oligonucleotides. The absorbance ratio of 260 nm and 280 nm will give an estimate of the purity of the solution. High quantities of pure DNA and RNA solutions have OD 260/OD 280 values of 1.8 and 2.0, respectively. This method is not useful for small quantities of DNA or RNA (<1μg/ml).

2. Determination by Ethidium bromide fluorescence

The amount of DNA is a solution is proportional to the fluorescence emitted by ethidium bromide in that solution. Dilutions of an unknown DNA in the presence of 2 μg/ml ethidium bromide are compared to dilutions of a known amount of a standard DNA solutions spotted on an agarose gel or electrophoresed in an agarose gel.

Restriction Enzymes

- Restriction and DNA modifying enzymes should be stored in an insulated cooler at

  -20°C in 50% glycerol in a non-frost free freezer.
- Care should be taken to allow the tubes to reach room temperature.
- While handling enzymes gloves should be worn as fingers contain nucleases.

- A new sterile pipet tip should be taken while working with restriction enzyme.
- The volume of the enzyme should be less than 1/10 of the final volume of the reaction mixture.

Chapter 8

Working with *E.coli*

Cell Genotypes

A large variety of bacterial strains, like *E. coli,* are frequently used in molecular biological experiments. Students should choose the strain or strains required for an experiment on the basis of genetic properties of the strains. It is thus pertinent that a sound knowledge of the genetic makeup of the *E. coli* stains is acquired.

The genetic makeup of the strains of *E.coli* is distinctly different from the wild type (K12 or B strains) strains from which they are derived and contains genes that can be modified. Genes are supposed to be mutated when they are not listed in a genotype. In most cases, the genes listed are mutated to the point of inactivity. In a few cases, a "-" (indicating inactive or absent or deleted), "+" (indicating active). , The genotype is a factor in choosing a bacterial strain. Chosen strains should be ascertained for their utility for the specific experiment.

Small Scale Cultures

Fresh culture from a freshly streaked plate should be taken while experimenting with *E. coli* cells. For preparing a small scale culture of

E. coli , prepare 3-5 ml of LB medium/broth (or appropriate broth - include antibiotic if the culture contains a plasmid) in two sterile 50 ml tubes. Inoculate one tube with a single colony from a fresh plate. The second tube is used as a broth control. Incubate both tubes at 37°C, shaking vigorously overnight. Inspect the tubes the next morning. The broth control should be clear and the inoculated culture should be very turbid. Make a note of any debris found in the tubes and only incubate longer if the culture is not dense. Never allow cells to overgrow. For some applications, cells can be stored at 4°C for short periods prior to use.

Permanent Storage

For every culture used, in particular, for newly constructed strains or for cells containing plasmids, a permanent glycerol stock must be prepared as soon as the construct has been confirmed. Also, all plasmid constructs, including construction intermediates, must be maintained in cells, not as naked DNA stocks. For each construct, at least 2 stocks should be made.

Preparation of glycerol stock for *E. coli* cells

Take 1.4 ml of a freshly grown overnight culture and mix it well with 0.6 ml of sterile 50% glycerol. Transfer to two freezer vials (not in eppendorfs) labeled with the strain name and date. Immediately place into a dry ice/ethanol bath or into a box in the -80°C freezer.

Sterile Technique

1. Autoclave all media, including plates, liquid media and top agar immediately after it is prepared. It is best to prepare media in several small bottles, only opening one at a time. Check the bottle for contamination before you use it by gently swirling it and looking for cloudy material in the center. Always grow up a small amount of broth alone when growing cells overnight. A small amount of contamination is not always evident until the media is incubated at 37°C.

2. Use a flame on inoculating loops and on the lips of media bottles before and after pipetting from them. Never leave a media or agar bottle open on the bench and do not take an individually-wrapped pipet out of its protective wrapper until you are ready to use it. Always use a fresh, sterile pipet or pipet tip when

pipetting culture media, and never go back into a media bottle or cell culture with a used pipet.

3. To prevent wide-scale, untraceable contamination, each person should have his own stock of liquid culture media, top agar, plates, 100% glycerol, glycerol stocks of cells, etc. and don't share.
4. Overnight cultures should be grown only from a single colony on a fresh plate or from a previously-tested glycerol stock that was grown from a single colony. To prepare an overnight culture from a glycerol stock, take an individually-wrapped 1-ml pipet and a culture tube of media to the -80°C freezer. Quickly remove the cap from the freezer vial containing the glycerol stock, scrap a small amount of ice from the surface of the culture, replace the cap on the freezer vial, and place the pipet into the culture tube. Sufficient numbers of bacteria are present in the ice in order for the culture to grow to saturation in 16 hours. Never let the glycerol stock thaw.

Colony Isolation

Step 1: Using a sterile loop, streak cultures (liquid broth or isolated colonies picked from plates) over one-fourth of the surface of an agar plate. Then flame the loop as described on the preceding page. Swabs containing an inoculum should be rolled over a small area of the agar surface to deposit the inoculum and then distributed over one-fourth of the surface with a sterile loop. The swab can then be used to prepare a stain or discarded.

Step 2: Air cool a flamed loop or cool it by touching an unstreaked area of agar on the same plate.

Step 3: Pass the cooled loop three or four times over the initial streaked portion of the plate. Streak it, without overlap, to the next quadrant.

Step 4: Flame the loop and allow it to cool as described above in Step 2.

Step 5: Pass the loop over the streaked portion of the second quadrant two or three times and then streak the material without overlapping over the third quadrant of the plate.

Step 6. Repeat Step 5 to streak the last quadrant.

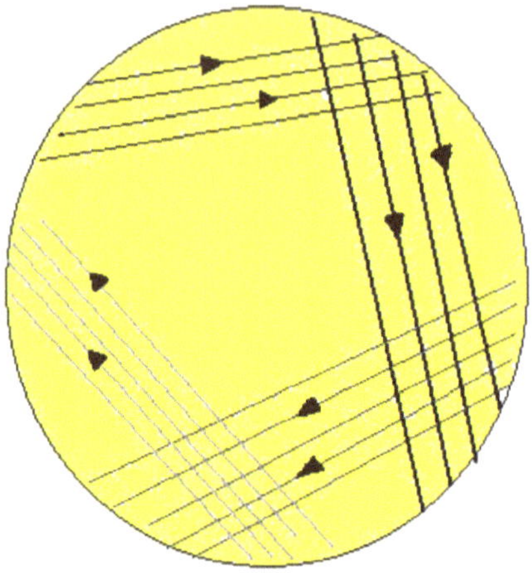

Fig.6: Method of streaking

Most bacteria do not move appreciably from the sites of inoculation but give rise to clones of bacteria called colonies. Isolated colonies should arise in the third and fourth quadrants depending on the concentration of bacteria in the initial inoculum.

Some of the criteria frequently used to characterize bacterial growth on agar media include:

- Colony size (using relative terms such as pinpoint, small, medium, large)
- Colony color
- Colony shape (describing the form, elevation, and edges of a colony)
- Colony surface appearance (e.g. dull, opaque, moist/glistening)
- Changes in agar media resulting from growth (e.g. hemolytic pattern on blood agar, changes in media or colony color due to a change in pH indicators, pitting of agar surface)
- Odor (certain bacteria produce distinct odors)

Chapter 9

Polymerase Chain Reaction

PCR Principles and Procedure

Polymerase chain reaction (PCR) is a technique that allows the generation of large amounts of a single DNA sequence from a mixture of sequences; the fragment generated can be designed to contain specific starting and ending positions based on the needs of the experiment. PCR is used to amplify specific regions of a DNA strand (the DNA target). This can be a single gene, a part of a gene, or a non-coding sequence. Most PCR methods typically amplify DNA fragments of up to 10 kilo base pairs (kb), although some techniques allow for amplification of fragments up to 40 kb in size.

A basic PCR set up requires several components and reagents. These components include:

- DNA template that contains the DNA region (target) to be amplified.
- Two primers that are complementary to the 3' (three prime) ends of each of the sense and anti-sense strand of the DNA target.

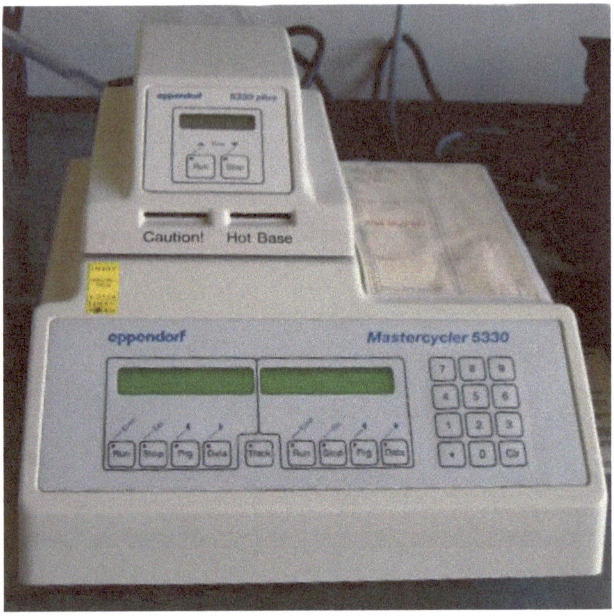

Fig.7: Picture of PCR Machine Thermal Cycler

- Taq polymerase or another DNA polymerase with a temperature optimum at around 70 °C.
- Deoxynucleoside triphosphates (dNTPs; also very commonly and erroneously called deoxynucleotide triphosphates), the building blocks from which the DNA polymerases synthesizes a new DNA strand.
- Buffer solution, providing a suitable chemical environment for optimum activity and stability of the DNA polymerase.
- Divalent cations, magnesium or manganese ions; generally Mg^{2+} is used, but Mn^{2+} can be utilized for PCR-mediated DNA mutagenesis, as higher Mn^{2+} concentration increases the error rate during DNA synthesis
- Monovalent cation potassium ions.

Components of the Reaction Mixture

1. Template DNA

Usually the amount of template DNA is in the range of 0.01-1 ng for plasmid or phage DNA and 0.1-1 µg for genomic DNA, for a total reaction mixture of 50 μl. Higher amounts of template DNA usually increase the yield of nonspecific PCR products, but if the fidelity of synthesis is crucial, maximal allowable template DNA quantities together with a limited number of PCR cycles should be used to increase the percentage of "correct" PCR products. Nearly all routine methods are suitable for template DNA purification. Although even trace amounts of agents used in DNA purification procedures (phenol, EDTA, Proteinase K, etc.) strongly inhibit *Taq* DNA polymerase, ethanol precipitation of DNA and repetitive treatments of DNA pellets with 70% ethanol is usually effective in removing traces of contaminants from the DNA sample

2. Primers

PCR uses an enzyme that synthesizes DNA (DNA polymerase). DNA polymerases require a "signal" to begin synthesis. The signal is a short fragment of DNA (known as primers). PCR uses two synthetic oligonucleotides (primers) that thus base-pair to the ends of the sequence of interest and act as initiation signals of DNA synthesis.

Logically the amount of DNA will double in each cycle and beginning with a single molecule of DNA it is possible to large amount of DNA. In the PCR reaction DNA strands will always have specific starting and ending points, where the primer sequences mark the ends of the fragments. Good PCR primers contain approximately 50% G+C and 50% A+T. This reduces problems in inducing template strand separation caused by the high affinity of G for C, and reduces non-specific priming common with high A-T content.

In addition, good primers have few regions of complementarity either internal to primer (*i.e.* secondary structure) or between two primers, especially at 3´ end of primer, and avoid repeated sequences (*e.g.*, AAAA or GTGTGT). Primers that are "poor" by these criteria can cause artifacts that prevent amplification of the desired sequence; hence the use of longer primers (*i.e.* = 24 bases) may solve specificity problems.

The primer should be long enough to have an optimum melting temperature (*i.e.* a melting temperature of 55-70°C), which depends on several factors. In most cases, an approximation [(4°C for each G or C) + (2°C for each A or T)] will yield a value close enough to design the PCR experiment.

Primer selection

- PCR primers are usually 15-30 nucleotides in length. Longer primers provide higher specificity.
- The GC content should be 40-60%. The C and G nucleotides should be distributed uniformly throughout of the primer. More than three G or C nucleotides at the 3'-end of the primer should be avoided, as nonspecific priming may occur.
- The primer should not be self-complementary or complementary to any other primer in the reaction mixture, in order to avoid primer-dimer and hairpin formation.
- The melting temperature of flanking primers should not differ by more than 5°C, so the GC content and length must be chosen accordingly.
- All possible sites of complementarity between primers and the template DNA should be noted.
- If primers are degenerate, at least 3 conservative nucleotides must be located at the primer's 3'-end.
- Estimation of the melting and annealing temperatures of primer:

If the primer is shorter than 25 nucleotides, the approx. melting temperature (Tm) is calculated using the following formula:

$$Tm = 4 (G + C) + 2 (A + T)$$

G, C, A, T - number of respective nucleotides in the primer.

$MgCl_2$ Concentration

Optimum concentration of $MgCl_2$ should be chosen as because Mg^{2+} ions complexes with

dNTPs, primers and DNA templates. Too little Mg^{2+} ions will result into a low yield of PCR product, and too many increase the yield of non-specific products and promote misincorporation. The optimum range of $MgCl_2$ concentration is 1-4 mM, under the standard reaction conditions. In an experiments, at a final dNTP concentration of 0.2 mM $MgCl_2$ concentration should range between 1.5±0.25 mM (in *Taq* buffer with KCl) and of 2.0±0.5 mM (in *Taq* buffer with $(NH_4)_2SO_4$). $MgCl_2$ concentration in the reaction mixture should be raised if the DNA samples contain EDTA or other chelators.

Table 3 : For selection of 25 mM $MgCl_2$ solution volume:

Final concentration of $MgCl_2$ in 50 µl reaction mix, mM	1.0	1.25	1.5	1.75	2.0	2.5	3.0	4.0
Volume of 25 Mm $MgCl_2$, µl	2	2.5	3	3.5	4	5	6	8

dNTPs

- It is very important to have equal concentrations of each dNTP (dATP, dCTP, dGTP, dTTP). The concentration of each dNTP in the reaction mixture is usually 200 µM. It is pertinent to maintain accuracy as even an additional dNTP increases the misincorporation level.
- The final dNTP concentration should be 10-50 µM. The fidelity of DNA synthesis is maximal in this concentration range, as maximum fidelity of the PCR process is very important. In addition, the concentration of $MgCl_2$ should be selected empirically, starting from 1 mM and increasing in 0.1 mM steps, until a sufficient yield of PCR product is obtained.

Taq DNA Polymerase

An amount of 1-1.5 u of *Taq* DNA polymerase is generally used in 50 µl of reaction mix. Synthesis of nonspecific products may result if higher amount of *Taq* DNA Polymerase is taken. However, in case there is presence of inhibitors in the reaction mix (like if the template DNA used is not highly purified), higher amounts of *Taq* DNA Polymerase (2-3 µ) will be necessary to obtain a better yield of amplified products.

Preparation of Reaction Mixture

It is essential to prepare a stock mix containing water, buffer, dNTPs, primers and *Taq* DNA polymerase in a single tube for performing several parallel reactions from which required amount can be aliquoted into individual tubes. $MgCl_2$ and template DNA solutions should then be added. This will minimizes the possibility of pipetting errors and saves time by reducing the number of reagent transfers.

Set Up of Reaction Mixture

1. Gently vortex and briefly centrifuge all solutions after thawing.
2. Add, in a thin-walled PCR tube, on ice:

Table 4: Reaction components

Reagent of reaction mixture	Final Concentration	Quantity, 50 µl
Sterile deionized water	-	variable
10X *Taq* buffer	1X	5 µl
2 mM dNTP mix	0.2 mM of each	5 µl
Primer I	0.1-1 µM	variable
Primer II	0.1-1 µM	variable
Taq DNA Polymerase	1.25 u / 50 µl	variable
25 mM $MgCl_2$	1-4 mM	variable
Template DNA	10pg-1 µg	variable

3. Gently vortex the sample and briefly centrifuge to collect all drops from walls of tube.
4. Overlay the sample with half volume of mineral oil or add an appropriate amount of wax. This step may be omitted if the thermal cycler is equipped with a heated lid.
5. Place samples in a thermocycler and start PCR.

Temperature Profile

The temperature profile shown below is a typical one that works for many different primers and templates:

Table 5 : PCR Temperature Profile

Temperature	Time	Function
94°C	0.5 minutes	Denature DNA
55°C	0.5 minutes	Primer Annealing
72°C	0.75 minutes	Primer extension

Steps in PCR

Initial Denaturation Step

- The complete denaturation of the DNA template at the start of the PCR reaction is of key importance. Incomplete denaturation of DNA results in the inefficient utilization of template in the first amplification cycle and in a poor yield of PCR product. The initial denaturation should be performed over an interval of 1-

3 min at 95°C if the GC content is 50% or less. This interval should be extended up to 10 min for GC-rich templates.

Cycle 1

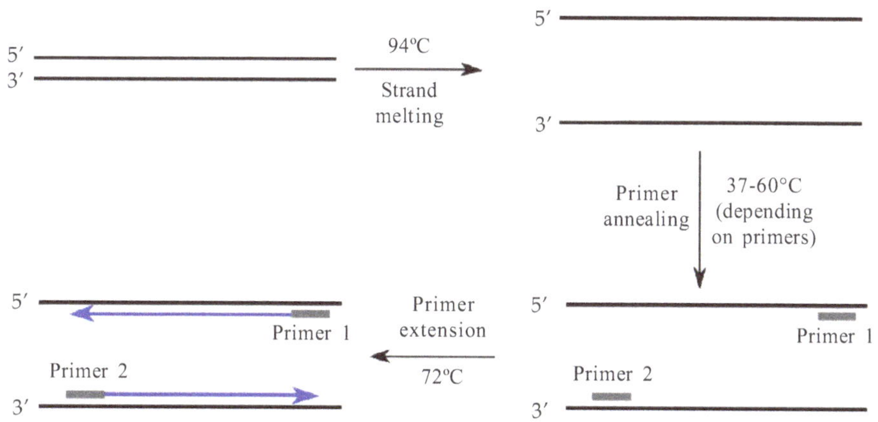

Fig.8: The two strands of the DNA template are separated by heating (usually to 94°C). The temperature is then decreased to allow the primers to bind to the template

Cycle 2+

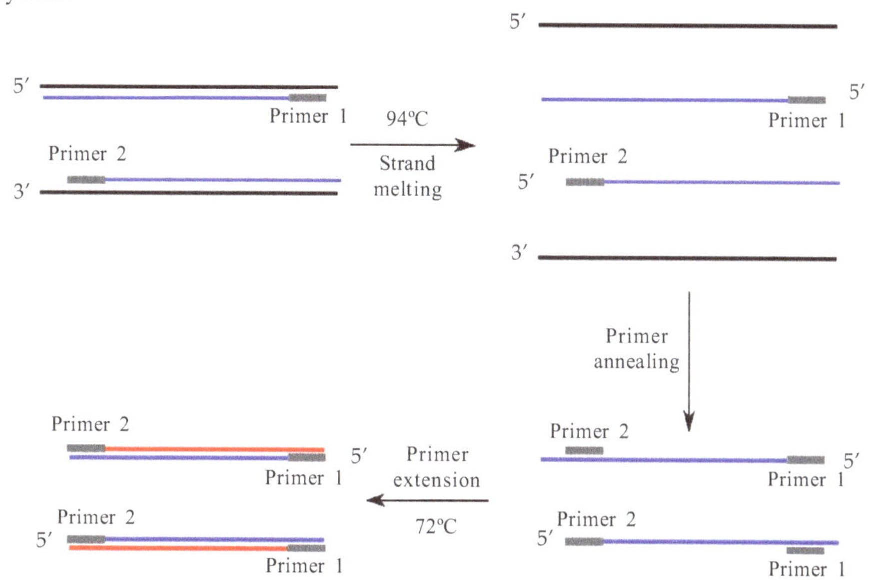

Fig. 9: Once the primers have bound, the polymerase is allowed to synthesize new DNA strands

- If the initial denaturation is no longer than 3 min at 95°C, *Taq* DNA polymerase can be added into the initial reaction mixture. If longer initial denaturation or a higher temperature is necessary, *Taq* DNA Polymerase should be added only after the initial denaturation, as the stability of the enzyme dramatically decreases at temperatures over 95°C.

Denaturation

The PCR product synthesized in the first amplification cycle is significantly shorter than the template DNA and is completely denatured under these conditions, hence it takes merely 0.5-2 min at 94-95°C for denaturation. Denaturation time increases up to 3-4 min. when the amplified DNA contains a very high GC content, Adding additives, for example: glycerol (up to 10-15vol.%), DMSO (up to 10%) or formamide (up to 5%), enhances DNA denaturation. In the presence of such additives, the annealing temperature should be adjusted as there is subsequent decrease in the melting temperature of the primer-template DNA duplex. Also the amount of enzyme in the reaction mix should be increased as DMSO and formamide inhibit *Taq* DNA polymerase by approx. 50%. A common way to decrease the melting temperature of the PCR product is to substitute dGTP with 7-deaza-dGTP in the reaction mix.

Primer Annealing

Usually the optimal annealing temperature is 5°C lower than the melting temperature of primer-template DNA duplex. Incubation for 0.5-2 min is optimum.

Extending Step

The extending step normally takes place at 70-75°C, because at this temperature the rate of DNA synthesis by *Taq* DNA polymerase is highest. The extension time for the synthesis of PCR fragments up to 2 kb is 1 min. The extending time should be increased by 1 min for each 1000 bp while amplifying larger DNA fragments.

Number of Cycles

The number of PCR cycles depends on the amount of template DNA in the reaction mix and on the expected yield of the PCR product. For less than 10 copies of template DNA, 40 cycles should be performed. If the initial quantity of template DNA is higher, 25-35 cycles are required.

Final Extending Step

After the last cycle samples should be incubated at 72°C for 5-15 min. to fill-in the protruding ends of newly synthesized PCR products. Also, during this step, the terminal transferase activity of *Taq* DNA Polymerase adds extra 'A' nucleotides to the 3'-ends of PCR products. Hence to clone a PCR fragment into T/A vectors, this step can be extended to up to 30 min.

Avoiding Contamination

PCR allows the production of more than 10 million copies of a target DNA sequence from only a few molecules. The sensitivity of this technique means that the sample should not be contaminated with any other DNA or previously amplified products (amplicons) that may reside in the laboratory environment.

- ❐ DNA sample preparation, reaction mixture assemblage and the PCR process, in addition to the subsequent reaction product analysis, should be performed in separate areas.
- ❐ A Laminar Flow Cabinet equipped with a UV lamp is recommended for preparing the reaction mixture.
- ❐ Fresh gloves should be worn for DNA purification and each reaction set-up.
- ❐ The use of dedicated vessels and positive displacement pipettes or tips with aerosol filters for both DNA sample and reaction mixture preparation, is strongly recommended.
- ❐ The reagents for PCR should be prepared separately and used solely for this purpose. Autoclaving of all solutions, except dNTPs, primers and *Taq* DNA Polymerase is recommended. Solutions should be aliquoted in small portions and stored in designated PCR areas. Aliquots should be stored separately from other DNA samples.
- ❐ A control reaction, omitting template DNA, should always be performed, to confirm the absence of contamination.

Chapter 10

Plasmid Preparation

Principle

Apart from PCR for generating large amounts of DNA, the second method is to use bacteria to replicate the DNA and then purify the replicated DNA from the bacteria. Plasmid DNA is usually the DNA of interest for this method. A plasmid is a double stranded circular DNA molecule that will replicate in an organism. A typical plasmid contains four important features:

- The plasmid must be circular, because bacteria generally will not replicate linear ↔ DNA.
- The plasmid must contain a sequence that functions as an origin of replication (ori).
- The plasmid must contain a selection mechanism that will force the bacteria to retain the DNA; the most common type of selection mechanism used in bacteria is a gene for resistance to an antibiotic such as ampicillin or tetracycline.

- The plasmid must contain a region for the insertion of the experimental DNA. A generic plasmid exhibiting these features is shown at right.
- An expression plasmid that allows expression of foreign DNA should have a strong promoter segment for transcription of inserted DNA.
- Must have an effective ribosome binding site that will allow efficient translation of the transcribed RNA.
- Must have high copy number, i.e., the host bacterium should have many copies (usually >100)

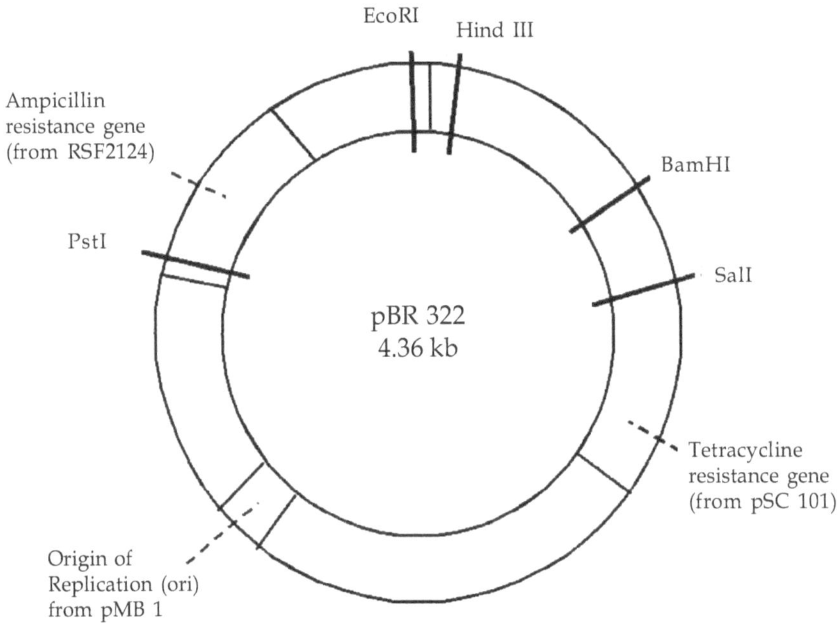

Fig. 10 : Plasmid pBR322

Plasmid preparation

1) Bacterial Growth with selected plasmid

The step involves culturing a single clone in a liquid culture (a single clone represent each colony on a plate). The bacteria are grown in the selected culture medium until they reach stationary phase. The stationary phase is defined as the stage when the bacteria either use up all available nutrients or when the bacterial waste products reach a concentration level that inhibit growth. In *E. coli*, the stationary phase reaches after 12-18 hours of growth (i.e., overnight growth result in achieving stationary phase). Some other bacterial types, and all yeast and mammalian cells in culture grow more slowly than does *E. coli*.

Note: The source of "foreign DNA" can be from any organism, including the host. Thus, *E. coli* can be used to express *E. coli* proteins from expression plasmids.

2) Separation of Bacteria from culture medium

This step fulfills two purposes:

- Concentrating the cells, and
- Transferring the cells into a buffer to facilitate plasmid purification.

This is done by centrifuging the culture and to discard the supernatant containing culture medium. The pellet containing bacterial cells can be either frozen for storage (this might promote cell lysis), or immediately transferred to a selected buffer for plasmid preparation.

3) Lysis of Bacterial cells to extract Plasmid DNA

Plasmid DNA extraction involves lysing the bacterial cells gently (to prevent damage of the released DNA) with a mixture of SDS and NaOH (as bacterial cells are much tougher than human cells), to disrupts the cell membranes and cell wall. The present alkaline lysis technique does not extract the chromosomal DNA as the intention of plasmid prep is only to purify the plasmid DNA while obtaining as little chromosomal DNA as possible.

4) Neutralization of alkaline pH

It is advisable to neutralize the alkaline pH imparted by NaOH once the cells have been lyses by adding an acetate buffer. Acetate buffer performs two important functions:

- Prevents degradation of the DNA, and
- Precipitates some proteins and most of the lipids.

5) Purification of Plasmid DNA

There are a number of techniques for separating plasmid DNA from the other molecules present in the cell. Most of these techniques involve binding the plasmid DNA to an insoluble material, washing off unbound material (protein, lipids, and small molecules), and then eluting the DNA.

Chapter 11

Plasmid DNA Isolation from *E.coli*

Isolation of plasmid DNA from *E. coli* is a common routine in research laboratories. One can perform a widely-practiced procedure that involves alkaline lysis of cells. This protocol often referred to as a plasmid "mini-prep," yields fairly clean DNA quickly and easily.

Procedure

1. Fill a microcentrifuge tube with saturated bacterial culture grown in LB broth + antibiotic. Spin tube in microcentrifuge for 1 minute, and make sure tubes are balanced in microcentrifuge. Dump supernatant and drain tube briefly on paper towel.
2. Repeat step 1 in the same tube, filling the tube again with more bacterial culture. The purpose of this step is to increase the starting volume of cells so that more plasmid DNA can be isolated per prep. Spin tube in microcentrifuge for 1 minute. Pour off supernatant and drain tube on paper towel.
3. Add 0.2 ml ice-cold Solution 1 to cell pellet and resuspend cells as much as possible using disposable transfer pipet.

- ❒ Solution 1 contains glucose, Tris, and EDTA. Glucose is added to increase the osmotic pressure outside the cells. Tris is a buffering agent used to maintain a constant pH (= 8.0). EDTA protects the DNA from degradative enzymes (called DNAse); EDTA binds divalent cations that are necessary for DNAse activity.

4. Add 0.4 ml Solution 2, cap tubes and invert five times gently. Let tubes sit at room temperature for 5 minutes.

- ❒ Solution 2 contains NaOH and SDS (a detergent). The alkaline mixture ruptures the cells, and the detergent breaks apart the lipid membrane and solubilizes cellular proteins. NaOH also denatures the DNA into single strands.

5. Add 0.3 ml ice-cold Solution 3, cap tubes and invert five times gently. Incubate tubes on ice for 10 minutes.

- ❒ Solution 3 contains a mixture of acetic acid and potassium acetate. The acetic acid neutralizes the pH, allowing the DNA strands to renature. The potassium acetate also precipitates the SDS from solution, along with the cellular debris. The *E. coli* chromosomal DNA, a partially renatured tangle at this step, is also trapped in the precipitate. The plasmid DNA remains in solution.

6. Centrifuge tubes for 5 minutes. Transfer supernatant to fresh microcentrifuge tube using clean disposable transfer pipet. Try to avoid taking any white precipitate during the transfer. It is okay to leave a little supernatant behind to avoid accidentally taking the precipitate.

- ❒ This fractionation step separates the plasmid DNA from the cellular debris and chromosomal DNA in the pellet.

7. Fill remainder of centrifuge tube with isopropanol. Let tube sit at room temperature for 2 minutes.

- ❒ Isopropanol effectively precipitates nucleic acids, but is much less effective with proteins. A quick precipitation can therefore purify DNA from protein contaminants.

8. Centrifuge tubes for 5 minutes. A milky pellet should be at the bottom of the tube. Pour off supernatant without dumping out the pellet. Drain tube on paper towel.

- ❒ This fractionation step further purifies the plasmid DNA from contaminants. This is also a good place to stop if class time is running out. Cap tubes and store in freezer until next class period.

9. Add 1 ml of ice-cold 70% ethanol. Cap tube and mix by inverting several times. Spin tubes for 1 minute. Pour off supernatant (be careful not to dump out pellet) and drain tube on paper towel.

❒ Ethanol helps to remove the remaining salts and SDS from the preparation.

10. Allow tube to dry for ~5 minutes. Add 50 µl TE to tube. If needed, centrifuge tube briefly to pool TE at bottom of tube. DNA is ready for use and can be stored indefinitely in the freezer.

Solutions

Solution 1: per 500 ml:	
50 mM glucose	9 ml 50% glucose
25 mM Tris-HCl pH 8.0	12.5 ml 1 M Tris-HCl pH 8.0
10 mM EDTA pH 8.0	10 ml 0.5 M EDTA pH 8.0
Add H_2O to 500 ml.	

Solution 2 : per 500 ml:	
1% SDS	50 ml 10% SDS
0.2 N NaOH	100 ml 1 N NaOH
Add H_2O	to 500 ml.

Solution 3:	**per 500 ml:**
3 M K^+	300 ml 5 M Potassium Acetate
5 M Acetate	57.5 ml glacial acetic acid
Add H_2O	to 500 ml.

TE	**per 100 ml:**
10 mM Tris-HCl pH 8.0	1 ml 1 M Tris-HCl pH 8.0
1 mM EDTA	0.5 ml 0.5 M EDTA pH 8.0
Add H_2O	to 100 ml.

Optional: DNase can be added to TE at final concentration of 20 µg/ml

Chapter 12

Plasmid Miniprep

Principle

The principle is to extract plasmid DNA from bacterial cell suspensions and is based on the alkaline lysis procedure developed by Birnboim and Doly (Nucleic Acids Research 7:1513, 1979). It is a known fact that plasmids are much smaller super coiled DNA molecules than the much larger and less super coiled bacterial chromosomal DNA. This difference in topology enables selective precipitation of the chromosomal DNA and cellular proteins and separates them from plasmids and RNA molecules. The cells are lysed under alkaline conditions, which denature both nucleic acids and proteins, and when the solution is neutralized by the addition of Potassium Acetate, chromosomal DNA and proteins precipitate because it is impossible for them to renature correctly (they are so large). Plasmids renature correctly and stay in solution, effectively separating them from chromosomal DNA and proteins.

Solution Preparation

Buffer 1

- 50 mM Tris-HCl,
- 10 mM EDTA,
- 100 μg/mL RNAse A, (pH 8.0)

Buffer 2

- 1% SDS,
- 0.2 M NaOH

Buffer 3

- 3.0 M Potassium Acetate, (pH 5.5)

TE buffer

- 10 mM Tris-HCl,
- 1 mM EDTA,
- pH 8.0

Procedure

The following procedure is for making duplicate minipreps, not only for balancing tubes for the centrifuge but also to acquire twice as much product when the experiment is finished.

1. Swirl gently the contents of the culture tube to resuspend the cells.
2. Two tubes of 1.5 mL should be labeled and 1000 μl of the cell suspension should be pipette into each tube.
3. The tubes should be capped and centrifuged at maximum speed for 20 s (always remember to balance the centrifuge by putting the tubes opposite one another). After some time the bacterial cells which are suspended in the liquid medium, settle into a cell pellet at the bottom of the tube.
4. The supernatant should be withdrawn and discarded using a pipettor. Care should be taken not to disturb the cell pellet.
5. 100 μl of Buffer 1 should be next added to each tube and the cells should be resuspended by vortexing. It's very important that the cell suspension is homogenous and no clumps are visible. The

cells are now resuspended in a buffered solution with RNAse. When the cells are lysed in the next step, the RNAse will catalyze hydrolysis of all RNA molecules into nucleotides, but the DNA will not be affected.

6. Next 200 µl of Buffer 2 should be added to each tube. The solutions should be then mixed by rapidly inverting them a few times after closing the caps. Avoid vortexing since the chromosomal DNA may get fragmented and contaminate the plasmid prep.
7. SDS (Sodium Dodecyl Sulfate), is an ionic detergent which disrupts cell membranes and destabilizes all hydrophobic bonds that hold various macromolecules in their native conformation. The high pH of the 0.2 M NaOH also denatures macromolecules by changing the condition of ionizable groups (ionizing certain groups and deionizing others).
8. Let tubes stand on ice for 5 minutes or longer.
9. Next 150 µl of ice-cold Buffer 3 should be added to each tube. The solutions should be then mixed by rapidly inverting them a few times after closing the caps. A white precipitate will form.
10. The low pH of the Potassium acetate solution neutralizes the NaOH, this renatures the macromolecules when the pH becomes neutral. The proteins and large DNA molecules do not renature correctly however. They form hydrophobic, ionic and hydrogen bonds with each other nonspecifically because the correct conformation of the molecule is not maintained during denaturation. The plasmid DNA molecules, however, never fully denatured because they are supercoiled small circular DNA molecules. They reform correctly when the pH is lowered even though the hydrogen bonds between base pairs were broken by the high pH. The large DNA molecules (chromosomal DNA) and proteins form precipitates because they bind to each other in a large. Hence, plasmids remain in solution while proteins and other larger DNA molecules precipitate.
11. The tubes should be kept on ice for 5 min or longer.
12. Next the tube should be centrifuged at maximum speed for 5 minutes. The precipitate will pellet along the side of the tube.
13. The supernatants should be next transferred into clean 1.5 mL tubes. Care should be taken to avoid sucking any precipitate. Discard the tubes with the precipitate.

14. An equal volume (about 400 μl) of isopropanol should be added next to precipitate the nucleic acids to each tube of supernatant.

    The tubes should be mixed thoroughly after closing the caps. After keeping the tubes at room temperature, with the hinges outward, centrifuge at full speed for 5 minutes. This step pellets the nucleic acids. Avoid keeping the tubes for too long as proteins in solution will begin to precipitate as well.

    This concentration of isopropanol will cause the DNA to become insoluble. Since DNA molecules are ionic (uniform negative charge due to the phosphates) they are highly soluble in water but not soluble in organic solvents. Isopropanol and ethanol are commonly used to precipitate DNA from aqueous solution.

15. The tubes with the hinges are kept outward because the plasmid DNA pellet may be difficult to see. It will be at the bottom and along the hinge side of the tube. Discard the supernatant carefully. The pellet is usually visible at this point. Although even if it is not visible yet there is probably enough DNA.

16. 200 μl of absolute ethanol should be added to each tube and mixed by inversion several times.

17. Centrifuge at maximum speed for 2-3 minutes (hinges out). Absolute Ethanol evaporates much faster than a solution of 50% isopropanol. The importance of this "ethanol wash" will speed up the process because the next step is to evaporate the alcohol used in the precipitation step.

18. The supernatant should be next removed and discarded carefully, without dislodging the pellet of plasmid DNA.

19. The last traces of ethanol should be dried off by keeping the uncapped tube at room temperature for 15-20 minutes.

20. 20 μl of TE buffer should be added next after last trace of ethanol is gone (you can check this by smelling the tube) to dissolve the pellet. The entire 20 μl should be pipette out.

21. TE buffer is commonly used to redissolve DNA because it contains EDTA. The EDTA will chelate magnesium ions which are a cofactor for most nucleases (enzymes which degrade nucleic

acids). If the DNA prep becomes contaminated with a nuclease (like the ones produced by the skin cells) then the nuclease will be inactivated because the magnesium cofactor is unavailable in the solution (because it is chelated by the EDTA).

22. Finally the contents of the two tubes should be pooled and kept in the freezer.

Chapter 13

Plasmid Midi-Prep from Bacteria

Solution Preparation

1. Glycerol mix (1 L)

- ❒ Weigh 25 grams (liquid weight) of glycerol and
- ❒ Add dH_2O to **1** Liter (Autoclave).

2. Potassium mix (1 L)

- ❒ 170 mL of 1M KH_2PO_4 (monobasic),
- ❒ 720 mL of 1M K_2HPO_4 (dibasic), and
- ❒ 110 mL of dH_2O (Autoclave)

3. $TE_{50/10}$ (100 mL)

- ❒ 5 mL of 1M Tris pH 7.5,
- ❒ 2 mL of 0.5M EDTA pH 8.0, and
- ❒ 93 mL of dH_2O

4. $TE_{10/1}$ (100 mL)

- ❒ 1.0 mL of 1M Tris pH 7.5,
- ❒ 200 µL of 0.5M EDTA pH 8.0, and
- ❒ 99 mL of dH_2O.

5. 4M LiCl (100 mL)

- 17 g of LiCl with
- 100 mL of dH_2O (Autoclave)

6. TB media (1 L)

- 12 g of Bacto Tryptone,
- 24 g of Bacto Yeast Extract, and
- dH_2O to 900 mL. Autoclave
- add 100 mL of Glycerol mix (after autoclave)

7. Lysis solution (30 mL)

- 1.5 mL of 20% SDS,
- 1.2 mL of 5M NaOH, and
- 27.3 mL of dH_2O.

8. 3M NaOAc pH 4.8 (100 mL)

- 29.4 g of potassium acetate,
- 40 mL of dH_2O,
- add Glacial acetic acid to pH 4.8, and
- dH_2O to 100 mL (Filter sterilize) (store at 4°C).

9. 1M KH_2PO_4 (100 mL)

- 13.6 g of KH_2PO_4 (monobasic), and
- 100 mL dH_2O (Autoclave)1

10. 1M K_2HPO_4 (100 mL)

- 17.42 g of K_2HPO_4 (dibasic), and
- 100 mL dH_2O (Autoclave)

Procedure

1. In a 125 mL flask, add: 36 mL of TB media + 4 mL of Potassium mix + applicable antibiotic
2. Transfer a single colony to the media and grow overnight
3. Centrifuge ~20 mL of culture at 6,000 rpm for 10 min. (4°C)
4. Decant supernatant and resuspend pellet in 1.5 mL of $TE_{50/10}$ + 4 μL of 21 mg mL^{-1} RNAse

5. Incubate on ice for 5 min. and add 3 mL of Lysis solution
6. Incubate on ice for 10 min. and add 2.25 mL of 3 M KOAc pH 4.8
7. Incubate on ice for 10 min. and centrifuge at 10,000 rpm for 15 min. (4°C)
8. Transfer supernatant to new tube containing 3 mL of 2-propanol and invert to mix
9. Centrifuge at 10,000 rpm for 30 minutes and discard supernatant
10. Let pellet drain and resuspend in 300 µL $TE_{10/1}$. Transfer solution to a 1.5 mL tube
11. Add 300 µL 4M LiCl and incubate on ice for 15 min.
12. Centrifuge for 15 min. at full speed (4°C) and transfer supernatant to a 2 mL tube
13. Fill the tube with 100% EtOH (~1.5 mL), mix by inversion (or: leave at 4°C ON) and centrifuge for 15 min. at full speed
14. Resuspend in 400 µL of $TE_{10/1}$ + 5 µL of 21 mg mL^{-1} RNAse and incubate for ~30 min. at 37°C
15. Extract with phenol: add 400 µL of TE-phenol, mix by vortexing, warm at 37°C for ~1 min, mix again, and centrifuge at full speed for 3 min.
16. Transfer upper transparent phase to a 1.5 mL tube, add 400 µL of 24 Chloroform: 1 isoamyl alcohol and carefully mix by vortexing
17. Centrifuge at full speed for 20 secs and transfer clear upper phase to a 1.5 mL tube
18. Add 40 µL of 3M NaOAc (Sodium Acetate) (1/10 vol) and ~1 mL of 100% EtOH (Ethidium Hydroxide)to the top of the tube and mix by inversion.
19. Centrifuge at top speed for 15 mins and aspirate off all the supernatant
20. Dry pellet for 5 mines.
21. Resuspend in 400-500 µL $TE_{10/1}$. The DNA sample is ready for downstream applications.

22. Determine concentration spectrophotometrically (~20 mL of culture should produce ~0.5-1.0 mg of plasmid DNA from a high copy plasmid)
23. One can determine the amount that is barely visible on a Thin EtBr-agarose gel, thicker gel mask the bands.

Chapter 14

DNA Plasmid Maxiprep

Preparation of Solutions

GTE buffer

- 50 mM Glucose,
- 25 mM Tris-Cl,
- 10 mM EDTA,
- pH 8

NaOH/SDS lysis solution

- 0.2 M NaOH,
- 1% SDS

TE buffer

- 10 mM Tris-Cl,
- 1 mM EDTA,
- pH 7.5

PEG solution

- 30% polyethylene glycol,
- 1.6 M NaCl

Procedure

1. Pick single colony and inoculate 250 ml of LB broth containing 100 μg/ml ampicillin or appropriate antibiotic. Shake at 250 RPM overnight.
2. Centrifuge cells at 5 k × g for 10 minutes.
3. Resuspend cell pellet in 10 ml of GTE buffer and add a spatula tip of lysozyme powder. A good suspension is consistent without clumps of cell pellet.
4. Add 20 ml of NaOH/SDS lysis solution. Use a spatula to stir and dissolve the cells until the solution becomes clear, yellow. This takes about 2 to 3 minutes and can be conducted at room temperature.
5. Add 15 ml of 5 M potassium acetate solution (pH 4.8). This solution neutralizes NaOH in the previous lysis step while precipitating the genomic DNA and SDS in an insoluble white, rubbery precipitate. Sir and vortex to neutralize. Centrifuge at 10 k × g for 10 minutes.
6. Into a clean bottle pour the supernatant through a small two-ply square of cheese cloth placed in the center of a funnel. The cheese cloth catches any fragments of SDS/genomic DNA pellet that are floating on the surface.
7. Precipitate the nucleic acids with 30 ml of isopropanol on ice for 10 minutes and centrifuge at 10 k × g for 10 minutes.
8. Aspirate off all the isopropanol supernatant. Dissolve the pellet in 5 ml of TE buffer. Transfer to a clean tube. Some nucleic acid pellet will be stuck on the sides of the bottle; thoroughly dissolve the entire pellet.
9. Add 5 ml of 5 M LiCl solution to precipitate RNA. Leave on ice for 10 minutes and centrifuge at 10 k × g for 10 minutes.
10. Pour off the supernatant containing plasmid DNA into a clean tube. Add an equal volume of isopropanol (10 ml) and precipitate

the nucleic acids on ice for 10 minutes. Centrifuge at 10 k × g for 10 minutes.

11. Aspirate off all the isopropanol supernatant. Dissolve the pellet in 1 ml of TE buffer. Transfer TE solution into a 1.5 ml eppendorfs tube. Add 15 µl of RNAse A solution (20 mg/ml stock stored at -20°C), vortex and incubate at 37 °C for 20 to 30 minutes to digest remaining RNA.
12. Precipitate the plasmid DNA with PEG solution by adding 0.4 ml and incubating overnight on ice for full recovery. This step discriminates very large plasmid DNA from small nucleic acid fragments as only the larger plasmid DNA precipitate.
13. Aspirate off the supernatant PEG buffer and dissolve the PEG pellet in 0.4 ml of TE buffer.
14. Extract proteins from the plasmid DNA using PCIA (phenol/ chloroform/isoamyl alcohol) by adding about 0.3 ml. Vortex vigorously for 30 seconds. Centrifuge at full speed for 5 minutes at room temperature. Note organic PCIA layer will be at the bottom of the tube.
15. Remove upper aqueous layer containing the plasmid DNA carefully avoiding the white precipitated protein layer above the PCIA layer, transferring to a clean 1.5 ml epindorf tube.
16. Add 100 µl of 7.5 M ammonium acetate solution and 1 ml of absolute ethanol to precipitate the plasmid DNA, usually on ice for 10 minutes. Centrifuge at full speed for 5 minutes at room temperature.
17. Aspirate off ethanol solution and resuspend or dissolve DNA pellet in 0.3 to 0.5 ml of TE buffer. This is the final stock of PEG pure plasmid DNA which is suitable for DNA sequencing and long term storage.
18. Measure the concentration of the plasmid DNA by diluting stock into water at 1:200 or 5 µl of DNA per 1 ml of water (blank spectrophotometer to water). The absorbance at 260 nm multiplied by ten is the concentration of the DNA in units of mg/ml for a 1 cm path length cuvette (i.e. 50 µg/ml/OD 260nm).

Chapter 15

Isolation of Genomic DNA from Bacteria

1. Transfer 1.5 mL overnight culture to a 1.5 mL microfuge tube, centrifuge for 30 sec, and decant supernatant.
2. Resuspend cells in 400 µL TE by vortexing, add 50 µL 10% SDS, 50 µL proteinase K (20 mg/mL in TE). Incubate for 1 hour at 37°C.
3. Shear DNA by 3-5 passages through a 26 G needle.
4. Extract twice with 500 µL phenol: chloroform (1:1), and twice with 500 µL chloroform.
5. Precipitate nucleic acids by adding 25 µL 5 M NaCl and 1 mL 95% EtOH, vortex, and centrifuge for 10 min, decant supernatant. Dry pellet.
6. Resuspend pellet in 100 µL TE buffer, add 5 µL RNaseA (5 mg/mL in TE), and incubate at 37°C for 30 min.
7. Precipitate DNA with 40 µL 5M NH_4Ac and 250 µL isopropanol; incubate at RT for 5 min.

8. Centrifuge for 10 min., wash pellet twice with 70% EtOH, dry pellet, dissolve in 100 μL TE.
9. Determine the DNA concentration by measuring the absorbance at 260 nm (1 OD_{260} = 50 μg/mL) or by comparison of the ethidium bromide staining intensity on a gel to a known amount of uncut lambda DNA.

Note: This procedure does not work well with Gram + cocci.

Chapter 16

DNA Preparation from Cell Lines, High Salt Method

Reagents

- ❐ EDTA
- ❐ Ethanol, absolute
- ❐ Isopropanol
- ❐ Phosphate Buffered Saline (PBS), 1X
- ❐ Proteinase K
- ❐ Sodium Acetate, 3 M pH 5.2 (Molecular Biology Grade)
- ❐ Sodium chloride (NaCl)
- ❐ Sodium Dodecyl Sulfate (SDS), 10%
- ❐ Tris EDTA (TE), pH 7.4
- ❐ Tris EDTA (TE), pH 8.0
- ❐ Tris HCl, pH 8.0

Preparation

Proteinase K solution

Proteinase K	100 mg
Tris EDTA, pH 7.4	10 ml

Nuclei lysis buffer

Tris EDTA, pH 8.0	1 ml
NaCl, 5 M	8 ml
EDTA, 0.5 M	0.4 ml
Add dH_2O to	100 ml

6 M Sodium chloride

Sodium chloride	3.5 g
dH2O	10 ml

Procedure

A. Cell Collection and Lysis

- ❐ Use trypsin or cell scraper to remove cells from tissue culture flask.Centrifuge cultured cells for 10 min at 1200 rpm.
- ❐ Remove and resuspend cell pellet twice with 10 ml 1X PBS, centrifuging between washes.
- ❐ Resuspend pellet in 10 ml nuclei lysis buffer. Centrifuge cultured cells for 10 min at 1200 rpm. Remove supernatant.
- ❐ Add 3 ml nuclei lysis buffer, resuspend the pellet, add 100 µl Proteinase K (10 mg/ml) and add 400 µl of 10% SDS, shake gently, and incubate at 45°C overnight.

B. Precipitation in High Salt Concentration

- ❐ To lysate, add 1 ml of 6 M NaCl
- ❐ Shake tubes vigorously by hand for 15 sec.
- ❐ Centrifuge at 3000 rpm for 15 min.
- ❐ Transfer supernatant into a new tube and centrifuge at 3000 rpm for 15 min.
- ❐ Repeat steps 3 and 4 until tube is clear of salt (at least 3-4 times).

C. Precipitation with Ethanol

- ❐ Transfer supernatant into a new tube; measure the volume of the supernatant.
- ❐ Add 1/10 the total volume 3 M sodium acetate (pH 5.2) and 2.5-3 times total volume cold 100% isopropanol; shake gently until the DNA is precipitated.

- ❐ Using the hook, transfer the DNA into a new tube containing 13 ml of 70% ethanol.
- ❐ Place on inverting rack and invert for 2 hrs to thoroughly rinse.
- ❐ Transfer DNA into new Eppendorf tube (1.5 ml) and centrifuge for 30 min at 14,000 rpm.
- ❐ Dry pellet by inverting on paper for 5 min.
- ❐ Add 200 µl dH20 and resuspend at 37°C overnight in Thermomixer.
- ❐ Measure the DNA concentration and run 1-5 µl (approximately 200 ng) for gel electrophoresis on agarose gel (1%) in 1X TAE buffer. Also, measure the DNA with Nano Drop and print out results for future reference.

Chapter 17

DNA Preparation from Fresh/ Frozen Tissues

Reagents

- ❐ Chloroform
- ❐ EDTA, 0.5 M
- ❐ Ethanol, absolute
- ❐ Isoamyl alcohol
- ❐ Phenol
- ❐ Phosphate Buffered Saline (PBS), 1X
- ❐ Proteinase K
- ❐ RNAse A
- ❐ Sodium dodecyl sulfate (SDS) solution, 10%

Preparation

DNA buffer (Tris-EDTA)

1 M Tris pH 8.0	20 ml
0.5 M EDTA	20 ml
Sterile water	100 ml

Proteinase K (10mg/ml)

- Dissolve 100 mg Proteinase K in 10 ml TE for 30 min at room temperature (RT)
- Aliquot and store at –20°C

RNAse A (20 mg/ml)

- Dissolve 200 mg RNAse in 10 ml sterile water, boil for 15 min, and cool to RT.
- Aliquot and store at –20°C

Procedure

- Put 60-80 mg of tissue in a petridish with culture media and divide the tissue into two pieces.
- Put the tissue into two sterile 15 ml tubes and centrifuge for 2 min at 4°C at 1500 rpm.
- Remove the supernatant, and wash twice with 1 ml 1X PBS or DNA-buffer. (It is possible to store the pellet at -80°C; in that case, add 1 ml 1X PBS and resuspend the pellet. Use a cryo-tube and centrifuge at 1500 rpm for 2 min at 4°C. Remove the supernatant, and freeze the pellet.)
- Remove supernatant and resuspend the pellet in 2.06 ml DNA-buffer.
- Add 100 μl proteinase K (10 mg/ml) and 240 μl 10% SDS, shake gently, and incubate overnight at 45°C in a water bath.
- If there are still some tissue pieces visible, add proteinase K again, shake gently, and incubate for another 5 hr at 45°C.
- Add 2.4 ml of phenol, shake by hand for 5-10 min, and centrifuge at 3000 rpm for 5 min at 10°C
- Pipette the supernatant into a new tube, add 1.2 ml phenol, and 1.2 ml chloroform/isoamyl alcohol (24:1); shake by hand for 5-10 min, and centrifuge at 3000 rpm for 5 min at 10°C.
- Pipette the supernatant into a new tube, add 2.4 ml chloroform/isoamyl alcohol (24:1), shake by hand for 5-10 min, and centrifuge at 3000 rpm for 5 min at 10°C.

- Pipette the supernatant into a new tube, add 25 µl 3 M sodium acetate (pH 5.2) and 5 ml ethanol, and shake gently until the DNA precipitates.
- Take a glass pipette, heat it over a gas burner, and bend the end to a hook. Fish the DNA thread out of the solution using the hook and transfer DNA to a new tube.
- Wash the DNA in 70% ethanol and dry it in the speed vac.
- Dissolve the DNA in 0.5-1 ml sterile water overnight (or longer if necessary) at 4°C on a rotating shaker.
- Measure the DNA concentration in a spectrophotometer and run 200 ng on a 1% agarose gel.

Table: 6 Ready reckoners for the experiment

Tissue (mg)	5	10	15	20	40	60	80	100
				Volume in µl				
Total	400	800	1200	1800	3200	4800	6400	8000
DNA buffer	360	680	1020	1360	2700	4080	5440	6800
Proteinase	20	40	60	80	160	240	320	400
10% SDS	40	80	120	160	320	480	640	800

Chapter 18

DNA Preparation from Paraffin Tissue

Reagents

- Chloroform
- EDTA, 0.5 M
- Ethanol, absolute
- Is amyl alcohol
- Phenol
- Proteinase K
- RNase A
- Sodium acetate, pH 5.2
- Sodium chloride, 5 M
- Sodium thiocyanate (NaSCN), 1 M
- TE Buffer (Tris-EDTA), pH 7.4
- Tween 20
- Xylene

Preparation

Chloroform/Isoamyl alcohol 24:1

Chloroform	24 ml
Isoamyl alcohol	1 ml

DNA extraction buffer

- 5M NaCl 1.5 ml
- 0.5M EDTA 5.0 ml
- Tween 20 0.5 ml
- Fill up to 100 ml with sterile water.

Proteinase K (10 mg/ml)

- Dissolve 100 mg Proteinase K in 10 ml TE for 30 min at RT.
- Aliquot and store at –20°

RNAse A (20 mg/ml)

- Dissolve 200 mg RNAse A in 5 ml sterile water.
- Boil for 15 min.
- Cool to room temperature
- Aliquot and store at –20°.

Procedure

- Cut 50 μm slices of formalin-fixed and paraffin-embedded tumor samples.

  (Note: Cut 4 μm slices before and after each 50 μm slice for Haematoxylin-Eosin staining to insure that the tissue is still representative).
- Incubate in xylene at 45°C for 15 min.
- Centrifuge 10 min at 14,000 rpm.
- Pipet off supernatant.
- Repeat steps 2-4.
- Add 1 ml 100% ethanol to tissue pellet, vortex, centrifuge for 10 min at 14,000 rpm, and pipet off supernatant.

- ❐ Add 1 ml 90% ethanol to tissue pellet, vortex, centrifuge for 10 min at 14,000 rpm, and pipet off supernatant.
- ❐ Add 1 ml 70% ethanol to tissue pellet, vortex, and centrifuge for 10 min at 14,000 rpm, pipet off supernatant and dry pellet in speed vac.
- ❐ Resuspend pellet in 1 ml NaSCN (1 M) and incubate at 37°C overnight.
- ❐ Centrifuge for 10 min at 14,000 rpm; pipet off supernatant.
- ❐ Resuspend pellet in 400 µl of DNA extraction buffer.
- ❐ Add 5 µl RNAse (20 mg/ml) and incubate for 1 hr at 37°C (RNAse treatment is optional, paraffin material often does not contain large amounts of RNA).
- ❐ Add 40 µl of Proteinase K (10 mg/ml), vortex briefly, and incubate at 55°C overnight (if tissue is not completely dissolved, add additional proteinase K and continue incubating; tissue should be dissolved).
- ❐ Add 440 µl of phenol, shake vigorously by hand for 5 min, and centrifuge for 5 min at 8000 rpm.
- ❐ Pipet supernatant into a new tube, add a solution of 220 µl phenol plus 220 µl chloroform/isoamyl alcohol (24:1), shake vigorously by hand for 5 min, and centrifuge for 5 min at 8000 rpm.
- ❐ Pipet supernatant into a new tube and add 440 µl chloroform/isoamyl alcohol (24:1), shake vigorously by hand for 5 min, and centrifuge for 5 min at 8000rpm.
- ❐ Pipet supernatant into a new tube (2 ml eppendorf tube), add 1/10 volume of sodium acetate (pH 5.2), add 3 volumes of ice cold 100 % ethanol, and keep tube for 1 hr at -80°C or overnight at -20°C.
- ❐ Centrifuge for 30 min at 4°C.
- ❐ Remove and save supernatant (optional, in case you are not sure of the precipitation).
- ❐ Dry pellet in speed vac.
- ❐ Add 20-50 µl sterile H_2O (depending on the amount of DNA you expect, which is subjective to experience).

- Shake gently in thermomixer at 37°C for 2 hours (DNA should be dissolved, but if you have doubts put on a rotating shaker in the cold room overnight).
- Measure DNA concentration with a spectrophotometer and run around 200 ng on a 1% agarose gel.

Section-3

Enzymes

Chapter 19

Restriction Digestion

Principle

Restriction digestion is one of the most common reactions performed in molecular biology. Restriction endonucleases (RE) are also known as restriction enzymes, which recognize short, specific DNA sequences. There are three types of RE: Type-I, Type-II and Type-III. Type II RE, most commonly used for DNA analysis and genetic engineering, cleave double-stranded DNA (dsDNA) at specific sites within or adjacent to their recognition sequences. All Type II RE recognises a unique nucleotide sequence at which it cuts a DNA molecule and nowhere else. The recognition sequence is often a six base pair palindromic sequence (the top DNA strand from 5' to 3' is the same as the bottom DNA strand from 5' to 3'), but others recognize four or even eight base pair sequences. Many RE will not cut DNA that is methylated on one or both strands of the recognition site, although some require substrate methylation.

REs can also differ in the way they cut the DNA molecule. Some enzymes cut in the middle of the recognition sequence, resulting in a flush or "blunt end". Other enzymes cleave in a staggered fashion, resulting in DNA products that have short single-stranded overhangs (usually two or four nucleotides) at each end. These are often called "cohesive ends", as these single-stranded overhangs could potentially come together again through complementary base-pairing, as shown below:

Blunt ends

N-N-A-G-C-T-N-N-	N-N-A-G	C-T-N-N
N-N-T-C-G-A-N-N-	N-N-T-C	G-A-N-N

Alu I

Cohesive ends

N-N-G-A-A-T-T-C-N-N	N-N-G	A-A-T-T-C-N-N
N-N-C-T-T-A-A-G-N-N	N-N-C-T-T-A	G-N-N

EcoRI

Importance of RE

Restriction enzymes are used to generate a "fingerprint" of a particular DNA molecule. Because of the sequence specificity of restriction enzymes, these enzymes can cut DNA into discrete fragments which can be resolved by gel electrophoresis. This pattern of DNA fragments generates a "DNA fingerprint," and each DNA molecule has its own fingerprint. Other restriction enzymes can be used to further characterize a particular DNA molecule. The location of these restriction enzyme cleavage sites on the DNA molecule can be compiled to create a restriction enzyme map. These maps are very useful for identifying and characterizing a particular DNA plasmid or region.

Set up of a restriction enzyme reaction

A restriction enzyme reaction contains mainly three components:

- The DNA to be analyzed,
- A restriction enzyme, and
- A restriction enzyme buffer mix.

Each restriction enzyme functions in a particular buffer mix. This buffer mix, commonly supplied at a 10X concentration, contains a buffering agent (usually Tris) to maintain constant pH, salt (usually NaCl or KCl) to provide the correct ionic strength for the digest, and Mg^{++} (from $MgCl_2$) as a necessary cofactor for enzyme activity.

Commercially available restriction enzymes usually have activities at 10-20 units/µl. A "unit" is usually defined as the amount of enzyme needed to digest 1 µg of bacterial virus lambda DNA in 1 hour in a 50 µl reaction. Total time required for complete digestion is usually 1.5 hours, but can be prolonged. The samples, at the end of the digestion can be stored in the refrigerator until ready to use. Loading dye can be added to the DNA digests, and the samples can loaded onto a gel.

Restriction enzymes are commonly stabilized in 50% glycerol solution. Do not exceed 5% glycerol in final digest with the two enzymes. Glycerol concentrations >5% may lead to star activity

Characteristics of Restriction Enzymes and Digestion

- The majority of restriction enzymes work best at 37°C
- Those isolated from thermophilic bacteria require higher temperatures for maximal activity (e.g., *BstX I* and *BstZ I* work best at 50°C).
- Some work below 37°C like SmaI (25°C) and CspI (30°C).

Chapter 20

Restriction Digestion of l-DNA with *Hind* III or *Eco* RI

Principle

Since the entire DNA sequence of the bacteriophage is known, we know exactly where a restriction enzyme will cleave the viral DNA and the size of the DNA fragments generated. This restriction enzyme digest will create a pattern of specific DNA fragments that will serves as a "DNA ladder".

Reagents

- 37°C incubator or water bath
- Electrophoresis units and power supplies
- 0.8% agarose in 1X TAE (melted)
- 1X TAE for electrophoresis units
- Methylene blue solution
- Lambda DNA (0.5 µg/ µl)
- Deionized or distilled H_2O

- 10X React 2 restriction enzyme buffer
- 10X React 3 restriction enzyme buffer
- Eco RI (10 units/µl)
- Hind III (10 units/µl)

Equipments

- Microcentrifuge tubes
- Micropipettes and tips
- Staining trays
- Light boxes

A restriction endonuclease buffer

- 10 mM Tris-HCl pH 8.8
- 1-5 mM $MgCl_2$
- 60 mM KCl
- 15 mM beta -mercaptoethanol and
- 600 mu M deoxyribonucleoside triphosphates.

Set up the restriction enzyme reaction of bacteriophage λ DNA

Set up the restriction enzyme digest as shown below (listed in order of addition):

Table 7 : Reaction mixture

Lambda DNA (0.5 µg/µl)	6µl
Deionized or distilled H_2O	10µl
10X React 2(Hind III) or 10X React 3 (EcoRI)	2 µl
Hind III or Eco RI	2 µl
Total volume	20µl

- Pool the reagents at the bottom of the tube. The tubes can be tapped to collect the fluids at the bottom.
- Incubate reaction tubes for at least 1.5 hours at 37°C (in a water bath or incubator). Store reactions in the refrigerator.

Agarose Gel Electrophoresis

- ❒ Pool the liquid from the restriction enzyme reaction at the bottom of the tube. Add 4 μl 6X loading dye to each reaction tube. Be sure the loading dye mixes with the DNA sample at the bottom of the tube.
- ❒ Record your electrophoresis unit number on your data sheet. Set up the electrophoresis unit by inserting the black casting dams and placing a comb in the slot closest to the black electrode.
- ❒ Pour the melted agarose into gel deck.
- ❒ Remove black casting dams and the comb.
- ❒ Record on data sheet where you will load the DNA samples.
- ❒ Load the DNA samples in the wells near the black electrode.
- ❒ Pour the TAE solution into the side farthest from you samples, so that you gels is completely covered plus a little more.
- ❒ Connect electrophoresis unit to power supply. The DNA will migrate through the gel for ~30 minutes.

Methylene Blue Staining

- ❒ Once the purple dye has migrated approximately 2/3 of the gel, turn off the power and carefully remove the gel.
- ❒ The gel is very fragile; take care not to break it. You can remove the tray that you poured agarose on to and gently slide the gel into the staining tray. At this point you cannot see the DNA.
- ❒ Once you have placed your gel into the staining tray bring it to the staining station. Completely cover the gel with methylene blue and cover try with saran wrap. Stain overnight.

Analysis of Results

Viewing the gel: Pour the methylene blue back into the bottle and carefully place the gel onto a white light box. The gel is very fragile so take care to not break it.

- ❒ Diagram the DNA bands on the gel data sheet. Be as accurate as possible in drawing the bands.

- Compare the banding pattern with a restriction enzyme map of bacteriophage lambda. Which DNA bands match with the lambda map?
- Use semilog paper to draw a curve comparing DNA length with distance migrated. Use the Hind III digest to generate the standard curve, and determine the DNA length of the *Eco RI* DNA fragments, based on the *Hind III* standard curve.

Chapter 21

Digestion with a Single and Double Restriction Enzyme

Reagents

- Nuclease-Free Water: 14μl
- 10X Restriction Buffer: 2μl
- Acetylated BSA (1mg/ml): 2μl
- DNA (~1μg): 1μl
- Restriction Enzyme (10u): 1μl
- Final Volume: 20μl

Procedure

- Mix by pipetting and collect the contents at the bottom of the tube.
- Incubate at the appropriate temperature for the enzyme for 1–4 hours.
- Add 4μl of 6X Blue/Orange Loading Dye and analyze digested DNA by gel electrophoresis.

- Preparing an insert for transfer from one vector to another usually requires a double digest (digest with two different REs). If both restriction enzymes work in the same restriction enzyme buffer, the reaction is straightforward.
- Simply add 1μl of the second restriction enzyme and adjust the amount of water used.

Note

In case if you want to work with two enzymes with different optimum temperatures, you can use the sequential digest method (perform for the lower-temperature digest first, and then digest at the higher temperature second, an hour at each temperature). When working with an enzyme that requires a temperature above 37°C, evaporation of the reaction can lead to increased glycerol concentration, which can in turn lead to star activity. Evaporation can be avoided in such reactions by applying a few drops of molecular biology grade mineral oil above the reaction.

Chapter 22

Quick Restriction Digestion

Reagents

- Restriction enzymes of choice, such as BamH1 and EcoRI
- Restriction enzyme reaction buffer,
- 70 % Ethanol
- 100 % Ethanol
- 3 M Sodium acetate (pH 5.2)
- Distilled water
- Plasmid DNA isolated from bacteria, such as pGFP(R) and pBC(R)

Instruments required

- Microcentrifuge with adaptors for 600-μl tubes
- Micropipetter and tips
- PCR machine and the 600-μl tubes for use in the machine

Table 8: Reaction Mixture

Reagents	Sample	Control	Control
distilled water	20 µl	21 µl	21 µl
10 X restriction enzyme reaction buffer	3 µl	3 µl	3 µl
DNA (at a concentration of 0.5-2gµ/ µl)	5 µl	5 µl	5 µl
BamHI (10 µ/ µl)	1 µl	0 µl	1 µl
EcoRI (10 µ/ µl)	1 µl	1 µl	0 µl

Procedures

- Set up the following 3 reactions (total of 30 µl each):
- Mix gently and spin for 2 seconds in a microcentrifuge.
- Incubate at an appropriate temperature for the restriction enzyme (usually 37°C) in the PCR machine for 30 minutes.
- Add 3 µl of the sodium acetate and 90 µl of 100 % ethanol.
- Store the tubes at -20°C for 1 hour.
- Spin in a microcentrifuge at maximum speed for 15 minutes.
- Discard supernatant and add 400µl of 70 % ethanol.
- Spin in a microcentrifuge at maximum speed for 5 minutes.
- Discard the ethanol and add another 400 µl of 70 % ethanol.
- Spin in a microcentrifuge at maximum speed for 5 minutes.
- Withdraw and discard as much liquid (ethanol) as possible then air-dry the pellet.
- Resuspend the pellet in 20 µl of distilled water.
- Keep at 4°C or -20°C for storage and to be examined later by agarose gel electrophoresis or DNA fragment purification.

Note

DNA should be stored at -20°C, but repeated freeze-thaw should be avoided. So if it is needed frequently or in a couple of days, store at 4°C.

Chapter 23

Ligation Reaction

Principle of DNA Ligase Mechanism

The process of connecting two pieces of DNA together is called ligation, and is catalyzed by an enzyme called ligase. The ligase used most often in molecular biology is derived from the T4 bacteriophage, and uses ATP to supply the energy necessary for the reaction. In addition to ATP, T4 ligase requires DNA with a 5′-phosphate group and free 3′-hydroxyl groups.

Applications

- Ligation of blunt or cohesive-ended DNA fragments.
- Repair of nicks in double-stranded nucleic acids.

DNA ligation takes place by the transfer of an AMP first to a lysine side chain on DNA ligase and from the latter to the 5'-phosphate group of the substrate. When the phosphodiester linkage in DNA is formed the AMP unit is released, as shown in the following reaction:

E + ATP (or NAD$^+$)	$\Longleftrightarrow$	E-AMP + PP$_i$ (or NMN)
E-AMP + P-5′DNA	$\Longleftrightarrow$	E + AMP – P– 5′-DNA
DNA-3′ -OH + AMP – P -5′-DNA	$\Longleftrightarrow$	DNA-3′-O –P-5′-DNA + AMP
DNA-3′-OH + P -5′-DNA+ ATP (or NAD)	$\Longleftrightarrow$	DNA-3′-O–P-5′-DNA AMP+PPi (or NMN)

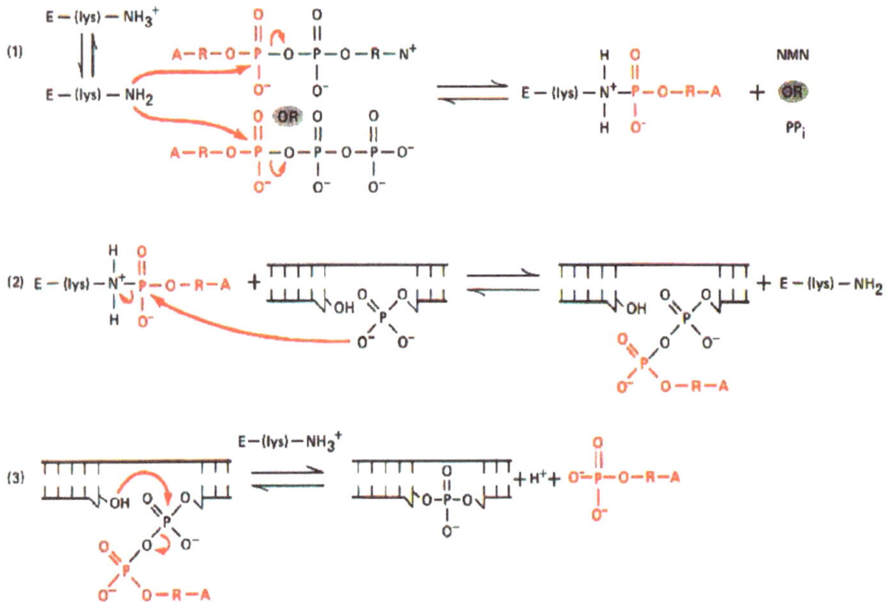

Synthetic oligonucleotides contain free 5′-hydroxyl groups, and therefore must be subjected to phosphorylation prior to ligation. In contrast, most restriction enzymes leave 5′-phosphate groups; most restriction fragments can be ligated immediately after digestion. Ligation also requires compatible ends to the DNA.

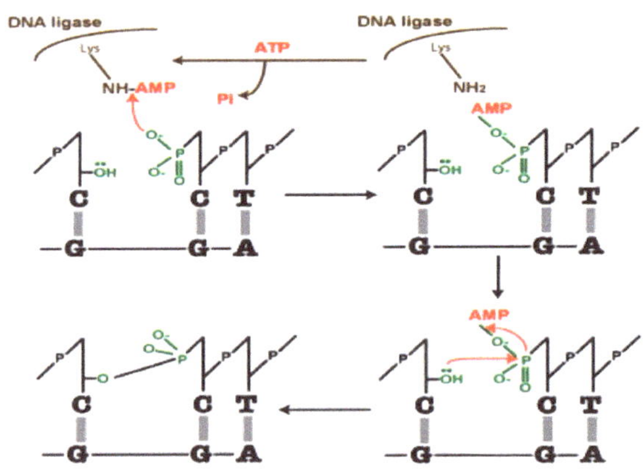

The reaction shown below is examples of different types of compatible and incompatible ends (N's can be any nucleotide present).

Compatible (Blunt Ends)		Compatible 4-base (5'Sticky Ends)		Compatible 2 base (3' sticky Ends)	
5'-NNNNN	NNNNN	5'-N	CCTGN	5'-NNNAC	NNN
3'-NNNNN	NNNNN	3'-NGGAC	N	3'-NNN	TGNNN

Incompatible (Blunt End/ non-blunt)	Incompatible 4-base (5'Sticky overhangs)		Incompatible 4 base (3' sticky overhangs)		
5'-NNNNN	N	5'-N	CCTGN	5'-NCTAG	N
3'-NNNNN	CATGN	3'-NGGAC	N	3'-N	ATTN

For example, if a plasmid is digested with two RE, *Nco* I and *Hin*dIII, it will create 4-base 5´-overhangs. The end generated by digestion with one of these enzymes is compatible with other ends generated by the same enzyme, but not by the ends generated by the other enzyme. Hence to construct a recombinant plasmid specificity of the restriction enzymes and ligase is important along with the insert in the correct orientation

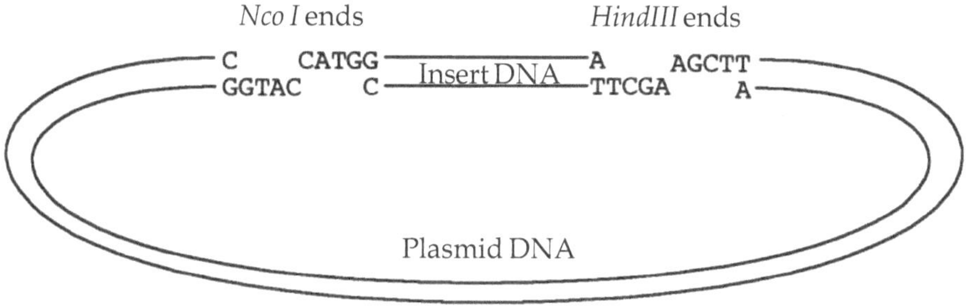

Fig.11: Plasmid with DNA insert

It is advisable to take an excess of insert DNA than the plasmid DNA in setting up a ligation reaction, without affecting the reaction. This is to prevent ligation of the plasmid DNA to form circular plasmid DNA and increases the chance of formation of recombinant DNA of plasmid DNA and insert DNA molecule.

Note: Ligation of incompatible ends, although very rare, does occur; circularized plasmid DNA will transform cells, and will result in colonies that lack insert.

Procedure of Ligation

- Mix cleaved plasmid DNA with a ~3- to 10-times molar excess of cleaved insert DNA.
- Add ligase buffer and ligase.
- Incubate at room temperature.
- After 2 to 24 hours, inactivate the ligase by heating at 68°C for 10 minutes.

T4 Ligase

T4 DNA Ligase is an ATP-dependent ligase used for DNA cloning. It joins dsDNA covalently with 5´-phosphorylated and 3´-hydroxylated blunt or compatible cohesive ends produced by RE digestion or other enzymatic processes. T4 DNA Ligase does not function on single-stranded nucleic acids. Following a ligation reaction, T4 DNA Ligase may be inactivated by incubation at 65°C for 10 minutes.

Ligation Reaction Buffer

- 32 mM Tris-HCl (pH= 7.6 at 25°C),
- 20 mM $MgCl_2$,
- 2 mM dithiothreitol,
- 2 mM ATP (to support the reaction),
- 15% Polyethylene glycol

Protocol

Combine 50 ng of vector with a 3-fold molar excess of insert. Adjust volume to 10 µl with dH_2O.

- Add 10 µl of 2X Quick Ligation Buffer and mix.
- Add 1µl of Quick T4 DNA Ligase and mix thoroughly.
- Centrifuge briefly and incubate at room temperature (25°C) for 5 minutes.
- Chill on ice, then transform or store at -20°C.
- Or may inactivate the enzyme by incubating at 65°C for 10 minutes.

Note: In case it is needed for transformations avoid inactivation by heating, which will otherwise reduce transformation efficiency.

Chapter 24

Fast Ligation of DNA

Start with a 1:2 molar ratio of vector: insert DNA when cloning a fragment into a plasmid vector. This ratio will vary with other types of vectors, for example, cDNA and genomic cloning vectors. The following example illustrates the conversion of molar ratios to mass ratios for a 3.0kb plasmid and a 0.5kb insert DNA fragment.

$$\frac{\text{Ng of vector x kb size of insert}}{\text{kb size of vector}} \times \frac{\text{molar ratio of insert}}{\text{vector}} = \text{ng of insert}$$

Example

How much 0.5kb insert DNA should be added to a ligation in which 100ng of 3kb vector will be used? The desired vector: insert ratio will be 1:2.

$$\frac{\text{100ng vector} \times \text{0.5kb insert}}{\text{3kb vector}} \times \frac{2}{1} = \text{33.3ng insert}$$

The following ligation reaction of a 3kb vector and a 0.5kb insert DNA uses a 1:2 vector:insert ratio. Typical ligation reactions use 100–200ng of vector DNA.

1. Assemble the following reaction in a sterile microcentrifuge tube:
   - Vector DNA 100ng
   - Insert DNA 33ng
   - 2X Rapid Ligation Buffer 5μl
   - T4 DNA Ligase 3μ
   - Nuclease-Free Water to final volume of 10μl
2. Incubate the reaction at room temperature for 5 minutes for cohesive-ended ligations, or 15 minutes for blunt-ended ligations.

   Notes:

   1. Ligation reactions performed using the 2X Rapid Ligation Buffer do not need to be cleaned up before transformation.

   2. Concatamers may form as ligation products. The extent of concatamer formation depends on the vector:insert ratio, incubation temperature and incubation time (1).This should be taken into account when screening transformants.

T4 DNA Ligase

- *Molecular Weight*: 68kDa.
- *Requirements*: Mg^{2+}, ATP and DTT). The optimum concentration of Mg^{2+} is0.5–1mM. Mn^{2+} may be substituted for Mg^{2+} but is only 25% as effective as $Mg2^{+}$.
- *Inhibition*: 50% inhibition by greater than 150mM NaCl (activity measured at nicks.
- Other inhibitors include 0.2M K^{+}, Cs^{+}, Li^{+}, NH_4^{+} and 1mM spermine.
- *Inactivation*: Heat at 70°C for 10 minutes.

Chapter 25

DNA Ligation with T4 DNA Ligase

Principle

The rate of joining of DNA fragments is affected by seven parameters:

- DNA concentration,
- Concentration of compatible ends,
- Size of the fragments,
- Presence or absence of complementary, single-stranded ends (some nucleases form blunt ends and in these cases special techniques are needed for joining),
- G + C content and length of the complementary single strands,
- Temperature of the reaction and
- Ionic composition of the solution.

The length and G + C content determine the strength of the hydrogen bonding between compatible sticky ends. The melting temperature (Tm) of AT pairs is much lower than that of GC pairs. In the cloning of *Eco*RI fragments of DNA into plasmid pBR325 the sequence of each single-stranded terminus is AATT; the value of Tm for the double-stranded joint formed by annealing these ends is 4°C-5°C. After allowing hydrogen bonds to form, the joint is covalently sealed with T4 DNA ligase, an enzyme whose activity is maximal at 37°C, a temperature at which the AATT double-stranded region would be disrupted. Fortunately, the enzyme does also function slowly at lower temperatures. At 12°C, the double-stranded AATT joint is very unstable, but not totally so, for it is being made and broken continually. Thus, if sufficient time is allowed at this temperature, the T4 DNA ligase succeeds in sealing the joints in the double-stranded segments that exist at a particular instant. Therefore, this temperature is usually selected as a compromise between rate and efficiency of annealing.

The configuration of single-stranded DNA is affected by ionic conditions. At high ionic strength single strands collapse and form intrastrand hydrogen bonds, which prevent intermolecular pairing. At low ionic strength single strands are extended, but electrostatic repulsion between the highly charged phosphate groups prevents the strands from coming into contact. Furthermore, owing to more subtle effects on the shape of double-stranded DNA, intramolecular joining is facilitated by high ionic strength. In some experiments, both intermolecular and intramolecular joining may be needed, the former to join different linear fragments and the latter for circularization of DNA prior to transformation. In the procedures used in these manual, moderately low concentrations of monovalent and divalent cations - 50mM and 10mM, respectively - are used. At these concentrations single strands are somewhat extended yet the charge is neutralized sufficiently to allow approach of the strands.

T4 DNA ligase will catalyze the ligation of either sticky or blunt ended DNA molecules with molecules of appropriate ends, but the reaction conditions under which the ligation optimally takes place are quite different depending on which type of end is being ligated. Unfortunately there is a wide disagreement in the literature as to how much ligase is needed, which buffer components to use and in what concentration. Use of a vast excess of ligase may actually be counterproductive as well as wasteful since the enzyme may be contaminated by trace amounts of nucleases which can damage the sticky ends and thus inhibit ligation. Some researchers advocate using a buffer which unlike the buffer in the protocol given here lacks ATP. This makes it easy to add whatever ATP concentration one wishes depending on whether one is doing blunt end or sticky end ligation. This is a useful point since many workers use a higher ATP concentration (1-5 mM) and/or add spermidine to 1 mM final concentration, particularly for sticky end ligations. High ATP and spermidine are inhibitory to blunt end ligation. The reason for this inhibition is unknown.

A. Sticky End Ligation

For ligations with sticky ends that are very AT rich (*e.g., Eco* RI) it may be desirable to lower the temperature slightly to 12°C -15°C, instead of 16°C. Whether or not ATP concentrations higher than 1 mM aid sticky end ligations is not clear. Blunt end ligations are essentially the same except that the amount of ligase in the reaction is 1-2 units rather than 0.1 units and the ATP concentration is 0.5 mM instead of 1 mM.

Procedure

1. Mix together vector and insert DNA at a 1:3 molar ratio of ends to achieve a total of 0.2 μg DNA.
2. *Add*:

- ❐ 2 μl 10x Ligase Buffer
- ❐ H_2O to 20 μl final
- ❐ Units T4 DNA Ligase

3. Incubate at 16°C overnight.
4. Transform *E. coli*.

Materials

1 M Tris, 7.6

- ❐ Compound Amount/ ml
- ❐ deionized H_2O 800 ml
- ❐ Tris base 121.1 g
- ❐ Adjust to desired pH by adding concentrated HCl (~ 50 ml).
- ❐ Allow the temperature to cool to room temperature before making the final adjustments to the pH.

1 M $MgCl_2$ (FW=203.3)

Compound Amount/ 1000 ml

- Deionized H_2O 800 ml
- $MgCl_2.6H_2O$ 203.3 g
- Dissolve and adjust to 1 L
- Dispense into aliquots and sterilize by autoclaving.

Note: $MgCl_2$ is extremely hygroscopic. Use small bottles (e.g., 100 g) and do not store opened bottles for long periods of time.

1 M Dithiothreitol (DTT, FW=154.3)

Compound Amount/ ml

- M Sodium acetate (pH 5.2) 20 ml
- Dithiothreitol 3.09
- Sterilize by filtration.
- Dispense into 1-ml aliquots and store -20°C.
- Do not autoclave DTT or solutions containing DDT.

10X Ligation Buffer

Final Concentration Stock Solution Amount/ 10 ml Amount/ 50 ml

- 0.5 M Tris, pH 7.6 1 M 5.0 ml 25 ml
- 100 mM $MgCl_2$ 1 M 1.0 ml 5 ml
- 100 mM Dithiothreitol 1 M 1.0 ml 5 ml
- 500 µg/ml Bovine Serum 5 mg 25 mg
- Albumin, Fraction V, Sigma (optional)
- Sterile deionized H_2O to volume

T4 DNA Ligase

16°C water bath

Sterile dH_2O

B. Blunt end Ligation with T4 DNA Ligase

Procedure

1. Mix together vector and insert DNA at a 1:3 molar ratio of ends to achieve a total of 0.2 µg DNA.

2. *Add:*
- ❒ 2 μl 10x Ligase Buffer
- ❒ H_2O to 20 μl final
- ❒ 2 Units T4 DNA Ligase
- ❒ Incubate at 20°C-26°C for 2-4 hours.

3. Transform *E. coli.*

   Note : The results of this protocol vary according to DNA purity, age of ligase, and the integrity of the blunt ends. Usual efficiencies are 5 x 104 transformants per μg of vector with a recombination efficiency of 10% assuming competent cells with an efficiency of about 5 x 105 colonies/μg of uncut DNA.

Materials

T4 DNA Ligase

1.0 or 5.0 mM ATP

dH_2O

100 mM Adenosine Triphosphates (ATP, FW=491.2)

- ❒ Compound Amount/1 ml Amount/10 ml
- ❒ dH_2O 800 μl 8 ml
- ❒ ATP 60 mg 600 mg
- ❒ Dissolve and adjust to pH 7.0 with 0.1 N NaOH
- ❒ dH_2O to 1 ml 10 ml
- ❒ Dispense the solution into small aliquots and store at -70°C.

Spermidine (50 mM)

- ❒ Compound Amount/ 10 ml
- ❒ Spermidine Dichloride (FW 254.63) 127.5 mg
- ❒ deionized H_2O 10 ml
- ❒ Mix in sterile tube, aliquot and store at -20°C.

Hexaminecolbalt chloride (10 mM, FW=267.5)

- ❒ Compound Amount/ 10 ml
- ❒ deionized H_2O 10 ml
- ❒ Hexaminecolbalt chloride, $Co(NH_3)_6Cl_3$ 26.8 mg
- ❒ Dissolve and filter sterilize

10X Blunt-End Ligation Buffer

Final Concentration Stock Solution Amount/ 10 ml Amount/ 50 ml

- 0.66 M Tris, pH 7.6 1 M 6.6 ml 33 ml
- 50 mM $MgCl_2$ 1 M 0.5 ml 2.5 ml
- 50 mM Dithiothreitol 1 M 0.5 ml 2.5 ml
- 1 mg/ml Bovine Serum 10 mg 50 mg
- Albumin, Fraction V, Sigma (optional)
- 10 mM hexamminecolbalt chloride 26.8 mg 134 mg
- 2 mM ATP 100 mM 0.2 ml 1 ml
- 5 mM Spermidine 50 mM 1.0 ml 5 ml
- Sterile deionized H_2O to volume. Store in small aliquot at -20°C.

Section-4

Cloning

Chapter 26

Competent Cell Preparation

Principle

Plasmids requires hosts to replicate, hence it has to be inserted into bacteria. But efficiency of uptake of plasmid DNA by bacteria is low and will not do so without getting degraded. Therefore, to enhance the probability of uptake of the plasmid DNA by bacteria it is necessary to make the bacterial cells "competent" to absorb the plasmid DNA. One of the methods to increase efficiency of cell competence is given below. It should be noted that competent cells are more fragile than normal bacteria. Vortexing the cells, heating the cells above 42°C or to 42°C for prolonged periods, or exposure of the cells to abusive treatments may kill them

Reagents Required

❒ LB medium	85%
❒ PEG(Polyethylene glycol) 8000	10%
❒ DMSO $(CH_3)_2SO$	5%
❒ $MgCl_2$	50 mM

- ❒ pH 6.5
- ❒ Sterile filter

LB medium

❒ Bacto-Tryptone	10g
❒ Bacto-yeast extract	5g
❒ NaCl	10g
❒ ddH_2O to	1litre
❒ Total volume	1litre

Note: Adjust pH to 7.0 and autoclave to sterilize.

Note: DMSO may degrade into transformation inhibitors; for best results, fresh DMSO should be stored in frozen aliquots.

Competent Cell Preparation

- ❒ Grow an overnight culture of the desired *E. coli* strain under the appropriate conditions (LB + antibiotic, if used).
- ❒ Add 1 ml aliquot of cells to 50 ml LB + 20 mM MgSO4; grow at ~18-20°C

  (Note: growth at higher temperatures results in somewhat lower efficiencies)
- ❒ Cool the cells on ice for 20 minutes.
- ❒ Spin down the cells (4000 rpm for 5 minutes) in sterile tubes.
- ❒ Resuspend the cells in 1/10 volume TSS (5 ml for 50 ml culture).
- ❒ Cells may be used immediately or frozen for storage.
- ❒ Quick freeze the cells in 310 µl aliquots in a dry ice: ethanol bath and store at -70°C.

Chapter 27

Cloning via Restriction Digests

Steps for Cloning

1. Prepare DNA.
2. Select restriction enzymes for your insert and vector, and determine the appropriate reaction buffers.
3. Combine the following in a microfuge tube (30 µL total volume):
   - ❐ 2 µg DNA
   - ❐ 1 µL Each Restriction Enzyme
   - ❐ 3 µL 10x Buffer
   - ❐ 3 µL 10x BSA
   - ❐ x µL H_2O (to bring total volume to 30 µL)

   Note: If you are using more than one restriction enzyme, depending on the buffers needed or your cloning strategy, you may need to digest with individual enzymes sequentially.
4. Incubate tubes at 37°C for 1 hour.

Note:

❐ *If you are doing a blunt-end ligation, use T4 DNA Polymerase or Klenow DNA Polymerase I for 3' overhang removal and 5' overhang fill-in. Use mung bean nuclease for both 3' and 5' overhang removal. Follow the manufacturer's instructions.*

❐ *If you are using blunt ends or a single enzyme to cut the vector, you will need to use a phosphatase to prevent re-circularization of the vector. CIP (calf alkaline phosphatase) or SAP (shrimp alkaline phosphatase) are commonly used. Follow the manufacturer's instructions.*

5. If you need to isolate your band of interest from other bands in the digest, proceed to Step 6. Otherwise, purify your DNA and proceed to Step 10 (ligation).

6. Prepare agarose gel. Low melting point agarose will facilitate DNA extraction at a later step, but regular agarose also works.

   ❐ To make a 0.8% agarose gel, use 0.8 gm agarose per 100 mL of 1x TAE. Cover the flask with plastic wrap to prevent boiling over, and then microwave the solution for 1-2 minutes to dissolve the agarose.

   Note: Increase the amount of agarose for better resolution of small bands and decrease the amount of agarose for better resolution of large band to ensure sufficient separation between your band of interest and neighboring bands.

   ❐ Let agarose solution cool for 5 minutes, and then add ethidium bromide to a final concentration of 0.5 μg/mL.

   Note: Run cold water over the outside of the flask for faster cooling.

   Safety tip: Ethidium bromide is a known mutagen. Wear a lab coat, eye protection, and gloves when working with this chemical.

   ❐ Pour the agarose/ethidium bromide solution into a casting tray with the well comb in place. Allow 20 -30 minutes to completely solidify.

   Note : Pour the gel in a 4°C room for faster solidifying. Gels can also be poured in advance, and stored in plastic wrap at 4°C.

7. Gel electrophoresis.

   ❐ Add 6 μL of 6x loading buffer to each of your samples.

   ❐ Place the agarose gel into the gel box (electrophoresis unit) and fill with 1xTAE until the gel is covered.

   ❐ Load a molecular weight marker into the first lane of the gel.

   ❐ Load your samples.

   ❐ Cover the gel box and plug in the electrodes. Black is negative, red is positive. (The DNA is negatively charged and will run towards the positive electrode.)

   ❐ Run the gel at 100V until the dye line is approximately 50-75% of the way down the gel.

   ❐ Carefully remove the gel from the gel box.

8. Extract DNA fragments.

   ❐ Visualize the bands under UV light. Use long-wavelength UV to minimize damage to the DNA.

   Note : When using UV light, protect your skin by wearing safety goggles or a face shield, gloves, and a lab coat.

   ❐ Using a clean razor blade, cut the desired DNA fragment from the gel. Turn the gel slice on its side to trim off extra agarose. Place the gel in a microcentrifuge tube.

9. Purification of DNA from the agarose gel:
   Run 1-2 μL of DNA on a gel to check the concentration of your insert and vector before proceeding.

10. Ligation:

❐ Combine the following in a microfuge tube (10 μL total volume):

❐ 1 μL Vector DNA

❐ 3 μL Insert DNA

❐ 1 μL 10x Ligase Buffer

❐ 1 μL T4 DNA Ligase

❐ 4 μL H_2O (to bring total volume to 10 μL)

❐ Incubate at 16°C for 2 hours, or at 4°C overnight

11. *Transformations*:

- ❒ Take competent cells out of -80°C and thaw on ice.
- ❒ Take agar plates out of 4°C to warm up to room temperature.
- ❒ Pipette 2 μL of the ligation and 30 μL of competent cells into each transformation tube.
- ❒ Leave on ice for 20 minutes.
- ❒ Heat shock each transformation tube at 42°C for 45 seconds.
- ❒ Leave on ice for 2 minutes.
- ❒ Add 1 mL LB and grow in 37°C shaker for 45 minutes.
- ❒ Spin down cells at 3000 rpm, and remove all but 100 μL of LB. Resuspend cells in the remaining LB.
- ❒ Spread onto an agar plate containing the appropriate antibiotic.
- ❒ Incubate plates at 37°C overnight.

Note: If your plasmid has ampicillin resistance and you do not require high transformation efficiencies, save time by skipping the recovery steps (steps 7-8) and directly spreading the 32 μL transformation mixture onto the agar plate.

Reagent

- ❒ Agarose
- ❒ Agarose, low melting point (LMP)
- ❒ DNA Ladder, 1 kb
- ❒ DNA Ladder, 100 bp
- ❒ DNA Polymerase, T4
- ❒ DNA Polymerase, Klenow
- ❒ Ethidium Bromide
- ❒ Ligase, T4
- ❒ Loading Buffer, 6X
- ❒ Mung bean nuclease
- ❒ Phosphatase, Calf Intestinal (CIP)

- ❐ Phosphatase, Shrimp Alkaline (SAP)
- ❐ Miniprep kit
- ❐ Gel Extraction Kit
- ❐ PCR purification kit
- ❐ Restriction enzyme
- ❐ TAE Buffer, 10X

Restriction analysis for plasmid verification

A restriction digest will allow you to verify the size of your plasmid. Also, it is a good idea to verify any restriction sites you will be using for cloning purposes.

1. Select restriction enzymes. Check the plasmid information page for insert and vector information.
2. Determine the appropriate reaction buffer. The optimal buffer is usually packaged with each enzyme during purchase. If you are conducting a double digest, check the enzyme's manufacturer for a table showing suggested buffers.
3. Combine the following in a microfuge tube (20 μL total volume):
   - ❐ 1 μg DNA
   - ❐ 0.5 μL Each Restriction Enzyme
   - ❐ 2 μL 10x Buffer
   - ❐ 2 μL 10x BSA
   - ❐ x μL H_2O (to bring total volume to 20 μL)
4. Incubate at 37°C for 1 hour.
5. *Prepare agarose gel:*

- ❐ To make a 0.8% agarose gel, use 0.8 g agarose per 100 mL 1x TAE. Cover the flask with plastic wrap to prevent boiling over, and then heat the solution for 1-2 minutes to dissolve the agarose.

  Note: Increase the amount of agarose for better resolution of small bands. Decrease the amount of agarose for better resolution of large bands.
- ❐ Let agarose solution cool for 5 minutes, and then add ethidium bromide to a final concentration of 0.5 μg/mL.

  Note: Run cold water over the outside of the flask for faster cooling.

Note: Ethidium bromide is a known mutagen. Wear a lab coat, eye protection, and gloves when working with this chemical.

❐ Pour the agarose/ethidium bromide solution into a casting tray with well comb in place. Allow 20-30 minutes to completely solidify.

Note: Pour the gel in a 4°C room for faster solidifying. Gels can also be poured in advance, and stored in plastic wrap at 4°C.

6. *Gel electrophoresis*:
   - ❐ Add 4 μL of 6x loading buffer to each of your samples.
   - ❐ Place the agarose gel into the gel box (electrophoresis unit) and fill with 1xTAE until the gel is covered.
   - ❐ Load a molecular weight marker into the first lane of the gel.
   - ❐ Load your samples.
   - ❐ Cover the gel box and plug in the electrodes. Black is negative, red is positive. The DNA is negatively charged and will run towards the positive electrode.
   - ❐ Run the gel at 100V until the dye line is approximately 50-75% of the way down the gel.
7. Carefully remove the gel from the gel box, and visualize the bands under UV light (254 nm).

   Note: When using UV light, protect your skin by wearing safety goggles or a face shield, gloves, and a lab coat.

Chapter 28

Transformation of Competent Cells

Transformation

The uptake of DNA by competent cells is called transformation. Transformation involves:

- Mix a small volume of DNA (usually 1-3 μl) with 100 μl of competent cells.
- The mixture is incubated on ice for 30 minutes to allow the cells to take up the DNA.
- The cells are then heat-shocked at 37 to 42°C for 1 minute, and cold shocked on ice for 2 minutes.
- The heat shock/cold shock reverses the effect of the competent cell process, by healing the damage inflicted by the competent cell solution.
- Add 0.5 ml of LB culture medium
- Incubate the cells for 10 - 30 minutes at 37°C to aid the healing process.
- This cell/medium mixture is then spread onto bacterial culture plates containing agar, LB, and a selection mechanism.

Chapter 29

Selection and Screening of Transformed Cells

Principle

Cloning experiments would be very difficult if we do not screen those very few cells that took up the recombinant plasmid DNA, as the transformation efficiency is very low as stated before (usually less than 1 in 10^5 cells). Hence it is necessary that there should be a selection mechanism (*i.e.* conditions where cells grow if they express a gene, but die if they do not) to sort out which are transformed cells and which are not make the job easy.

A commonly used selection mechanism is antibiotic resistance. If the *E. coli* strain used is not resistant to the antibiotic, but the plasmid DNA contains a gene coding for resistance, then only the cells that have taken up the plasmid will be able to grow. Most commonly used plasmids contain a gene for resistance to ampicillin; only cells that have taken up the plasmid will be able to grow and form colonies.

Note: Uptake of plasmid DNA is rare; hence each group of cells visible on a plate represents the offspring of a single cell that will contain a single molecule of plasmid DNA.

Screening the correct Plasmid

In most cloning procedures only a fraction of the colonies contain the correct plasmid. The selection technique mentioned above means that only cells containing plasmid DNA have grown; but it is still possible that some of the cells might have taken up incorrect plasmids (as long as the plasmid contains an ampicillin resistance gene and origin of replication). The figure below shows the starting plasmid (on the left), and a plasmid you might wish to construct (on the right). The next step, therefore, is the screening of the correct plasmid.

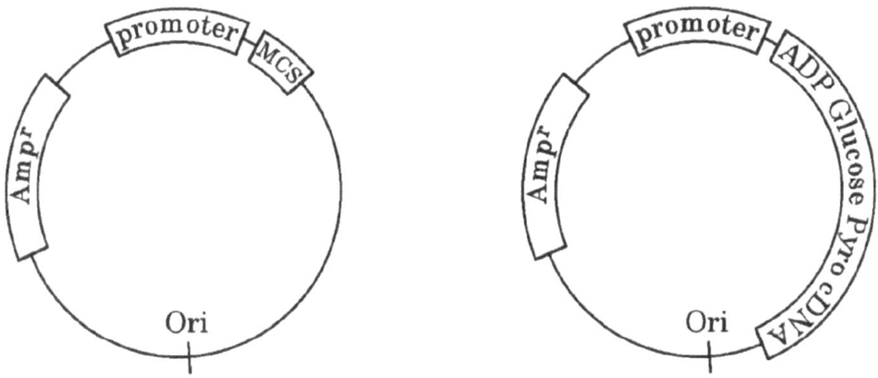

Fig.12: Plasmids with specific genes

One of the commonest method of screening of correct plasmid that might be found in cells on the same culture plate involve directly testing for the presence of specific DNA.There are a number of methods for this. Like:

- Performing PCR on the colonies to look for the presence of the specific insert DNA or
- Hybridization screening of the cells for the presence of the DNA looks directly at the DNA within the cells.
- Purifying the plasmid DNA and then testing the plasmid by restriction analysis.
- DNA sequencing also looks directly at the plasmid DNA.
- Screening technique can analyze the DNA indirectly, by looking for production of a heterologous protein or for an activity that must be due to a protein encoded by the plasmid of interest.

Among all the screening technique purifying plasmid DNA is much more labor intensive and expensive than those that assay the cells directly. Thus, PCR screening and hybridization screening (which are performed on cells derived directly from the culture plate) are preferable. PCR screening is best suited for detecting the presence or absence of relatively large DNA fragments.

PCR screening procedure

The procedure is to

- Pick the colony with a sterile pipet tip,
- Streak the cells on a new plate (so that, if the colony turns out to be correct, you will still have viable cells to replicate), and
- Next soak the pipet tip in 10 μl of water.
- This water (actually the few cells that drop off into the water) can then act as a template for a PCR.
- Colonies that result in PCR products of the correct size (by running the samples on an agarose gel) can be assumed to contain the correct plasmid.

A PCR screening reaction does not require a large reaction volume. A common method is to prepare a master solution of all of the required reagents except template, followed by adding the same volume of master mix to each PCR tube. The last step should be the addition of the template to the PCR tube (adding the template last reduces the chance of cross-contamination of the PCR samples).

A master mix that allows a final reaction volume of 25 μl and a template volume of 5 μl can be prepared as follows:

Master Mix (Per sample to be run):

1.25 mM dNTP mixture	4 μl
10x PCR buffer	2.5 μl
20 μM 5′ primer	1.25 μl
20 μM 3′ primer	1.25 μl
Thermostable DNA polymerase	0.125 μl
autoclaved water	11 μl

Section-5

DNA / RNA Separation

Chapter 30

Electrophoresis

Principle

The principle of Electrophoresis involves the motion of a particle in a fluid medium, where the gravitational attraction (on a flowing liquid) is the sedimenting force, and the frictional force is the opposing force and is proportional to the velocity of the particle. These forces balance each other acting in opposite directions and hence balance each other. This will lead to the uniform motion of the particle in the mobile liquid medium (i.e. the particle moves at constant velocity). In case the external field is an electric field instead of a gravitational field, a macromolecule will respond to the external field in two ways. In case the molecule is charged, it will migrate in an electric field to the electrode of opposite charge. Under this principle the technique of electrophoresis functions.

All electrophoresis operations are governed by the principle:

$$\text{Molecule a of mobility} = \frac{\text{(Voltage applied) (Molecule on the charge net)}}{\text{(molecule friction)}}$$

Electrophoresis is a process in which molecules are exposed to an electric field and separated on the basis of their differential mobilities in that field. The observed differential mobility is a result of different charge magnitudes on different molecules, and the result of different resistance to movement through the medium. For molecules with similar shapes, the mobility is proportional to the charge-to-mass ratio of the molecule. For molecules of similar shapes and similar charge-to-mass ratios, the motion through the medium will be proportional to the size of the molecule, because friction increases as a function of size.

The velocity of a charged molecule is given by:

$$v = \frac{qE}{f}$$

Where:

q is the charge on the molecule,

E is the electrical potential gradient, and

f is the frictional coefficient of the medium for the molecule.

For similarly shaped molecules, f is proportional to size of the molecule. For molecules in which charge increases in proportion to size, larger molecules move more slowly than small ones, because f increases faster than charge.

Modern day electrophoresis is conducted in solid gels (such as agarose or polyacrylamide). The solid gel is porous to solute and solvent molecules and serves as a medium for electrophoresis.

Chapter 31

Agarose Gel Electrophoresis

Agarose gel electrophoresis is the easiest and commonest way of separating, capable of resolving mixtures of DNA fragment that cannot be separated conveniently by other procedures. The purpose of the gel might be to look at the DNA, to quantify it or to isolate a particular band. The DNA is visualised in the gel by addition of ethidium bromide. This binds strongly to DNA by intercalating between the bases and is fluorescent meaning that it absorbs invisible UV light and transmits the energy as visible orange light.

Agarose, which is extracted from seaweed, is a linear polymer whose basic structure is shown in figure.

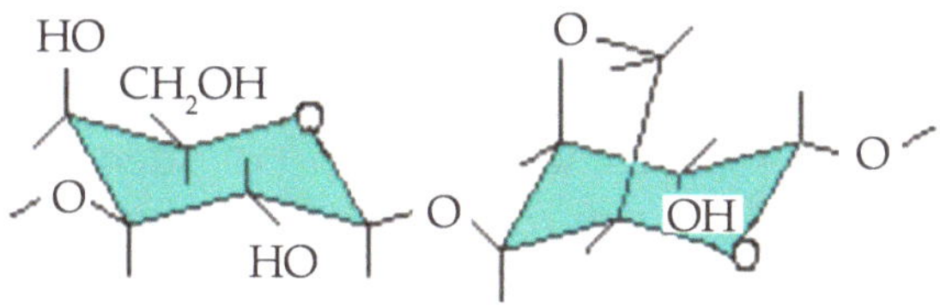

Fig.13: Structure of Agarose

Material

- DNA samples
- DNA size standard
- Loading dye solution
- Agarose
- Buffer/s
- Transilluminator
- Plastic wrap
- Micropipette

Preparation of Agarose Gel

DNA Agarose gels can be used to separate and visualize DNA of various sizes. Before casting an Agarose gel, the following table indicates how to determine the appropriate percent Agarose gel to use, based on the size of DNA to be separated.

Agarose Concentration

By using gels with different concentrations of agarose, one can resolve different sizes of DNA fragments. Higher concentrations of agarose facilitate separation of small DNAs, while low agarose concentrations allow resolution of larger DNAs.

Table 9: Range and Agarose concentration

Agarose concentration (%)	Separation range (Kb)
0.3	5 to 60
0.6	1 to 20
0.8	0.8 to 10
1.0	0.4 to 8
1.2	0.3 to 6
1.5	0.2 to 4
2.0	0.1 to 3

The figure below shows migration of a set of DNA fragments in three concentrations of agarose, all of which were in the same gel tray and electrophoresed at the same voltage and for identical times. Notice how the larger fragments are much better resolved in the 0.7% gel, while the small fragments separated best in 1.5% agarose. The 1000 bp fragment is indicated in each lane.

Procedure of preparing agarose gel

- Determine the amount of Agarose (grams) required to make the desired Agarose gel concentration and volume. Example: For a 1% Agarose gel, add 1 gram of Agarose to 100 ml of 1x electrophoresis buffer.
- Add the Agarose to a suitable container. Add the appropriate amount of 1x electrophoresis buffer and swirl to suspend the Agarose powder in the buffer.

  Note: A mark can be put on the flask at the same level as the liquid. If evaporation occurs, water can be added to bring the liquid back to the original starting level.
- Place the gel solution and boil and swirl the solution until all of the small translucent Agarose particles are dissolved. Taking out the flask from the flame swirling the flask every 30 seconds helps dissolve the Agarose faster.
- Cool the molten Agarose to 60 °C before pouring an Agarose gel slab.

  Note: 0.5 μg/ml ethidium bromide may be added to the molten Agarose. Alternative methods are described under Nucleic Acid Staining and Visualization.

Casting Agarose Gel

There are several ways to cast Agarose submarine gels using a removable tray with the aid of the gel caster or with standard laboratory tape.

Voltage Applied

As the voltage applied to a gel is increased, larger fragments migrate proportionally faster than small fragments. For that reason, the best resolution of fragments larger than about 2 kb is attained by applying no more than 5 volts per cm to the gel (the cm value is the distance between the two electrodes, not the length of the gel).

Nucleic Acid Electrophoresis Buffers

Several different buffers have been recommended for electrophoresis of DNA.DNA agarose gel electrophoresis is usually conducted with either Tris-Acetate-EDTA (TAE) or Tris-Boric Acid-EDTA (TBE). While TAE provides faster electrophoretic migration of linear DNA and better resolution of supercoiled DNA, TBE buffers have a stronger buffering capacity for longer or higher voltage electrophoresis runs. DNA fragments will migrate at somewhat different rates in these two buffers due to differences in ionic strength. Buffers not only establish a pH, but provide ions to support conductivity. If you mistakenly use water instead of buffer, there will be essentially no migration of DNA in the gel! Conversely, if you use concentrated buffer (e.g. a 10X stock solution), enough heat may be generated in the gel to melt it.

Buffers

There are a number of buffers used in this process, like

50x TAE (per litre)

- 242 gm Tris base,
- 57.1 gm glacial acetic acid,
- 100 ml 0.5 M EDTA
- pH 8.0

10x TBE (per litre)

- 108 g Tris base,
- 55 g boric acid,
- 40 ml 0.5 M EDTA,
- pH 8.0

20 x TTE buffer (per litre)

- Tris Base 216 gm,
- Taurine 72 gm,
- EDTA disodium salt 4 gm

6x gel loading buffer

- 0.25% Bromophenol blue,
- 0.25%Xylene cyanol FF,
- 15% Ficol Type 4000,
- 120 mM EDTA

DNA and RNA Sample Loading Dye

A convenient 10x sample buffer stock consists of:

- 50% glycerol,
- 0.25% bromophenol blue, and
- 0.25% xylene cyanole FF in 1x TAE buffer.
- Only 1–10 ml of the 10x loading dye should be prepared.

Setup and Run Electrophoresis

After the agarose gel has solidified, sample loading and electrophoresis can begin. Agarose gels can be run in many different types of electrophoresis buffers. Nucleic acid agarose gel electrophoresis is usually conducted with either Tris-Acetate-EDTA (TAE) buffer or Tris-Borate-EDTA (TBE) buffer. While TAE buffer provides faster electrophoretic migration of linear DNA and better resolution of supercoiled DNA, TBE buffers have a stronger buffering capacity for longer or higher voltage electrophoresis runs.

- Dissolve 1 gm of agarose in 100 ml of 1X TAE or TBE buffer (gives a 1% gel).
- Cast the gel with the comb in place
- Add 6X gel loading buffer to sample and load the samples into wells
- Run the gel submerged in 1X TAE or TBE (30-60 min. at 100-150V)
- Stain the gel with ethidium bromide solution (10 μl per 100ml of buffer) for 10 - 15 min.
- View on a UV transilluminator.
- Prepare samples for gel loading. Loading volume is dependent upon the type of comb used (*i.e.*, well thickness and length) and thickness of the gel.
- When loading volume is determined, add standard nucleic acid sample loading dye to a final 1x concentration to make samples dense for underlying into sample wells.
- Load the samples into the wells using standard pipettes.

Note: Sample wells are often difficult to see. Well visualization can be enhanced by placing black paper or tape under the base or trays where comb placement or well formation is common.

- Place the lid on the DNA cell carefully. Do not disturb the samples.

- Power requirements vary depending on gel thickness, length and concentration, and type of electrophoresis buffer used. Use 20V to 110 V for most purpose.

Gel Staining

There are different methods of staining nucleic acids for visualization using ethidium bromide. A fast and sensitive method which is extensively used (although due to safety concerns, it is going out of favor) is ethidium bromide (EtBr) staining. Three common applications include:

- Add 0.5 µg/ml directly to Agarose gel
- Add 0.4 µg/ml directly into the sample and loading on EtBr-free gel
- Stain the gel after completion of electrophoresis

  Note: Ethidium bromide is a suspected carcinogen and should be handled with extreme care. Always wear gloves, eye glasses, and a laboratory coat. Dispose of used EtBr solutions and gels appropriately.

Ethidium Bromide Staining Procedure after Electrophoresis

Ethidium bromide is a fluorescent dye that intercalates between bases of nucleic acids and allows very convenient detection of DNA fragments in gels, as shown by all the images on this page. As described above, it can be incorporated into agarose gels, or added to samples of DNA before loading to enable visualization of the fragments within the gel. As might be expected, binding of ethidium bromide to DNA alters its mass and rigidity, and therefore its mobility.

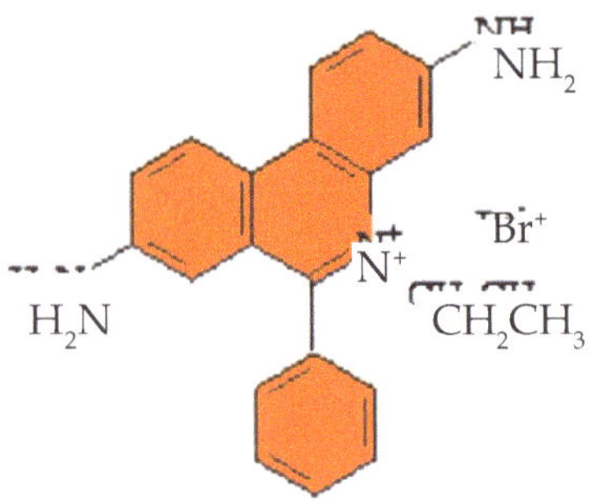

Fig. 14: Structure of Ethidium Bromide

- Place the gel into the appropriate volume of 0.5 µg/ml ethidium bromide (EtBr) stain for 15-30 minutes. Use enough staining solution to cover the entire gel.
- Destain the gel for 10-30 minutes in dH_2O using the same volume used for staining.

  Note: Ethidium Bromide can be removed from the DNA with extended destaining. This will cause lower sensitivity of detection. However, insufficient destaining will create higher background fluorescence.

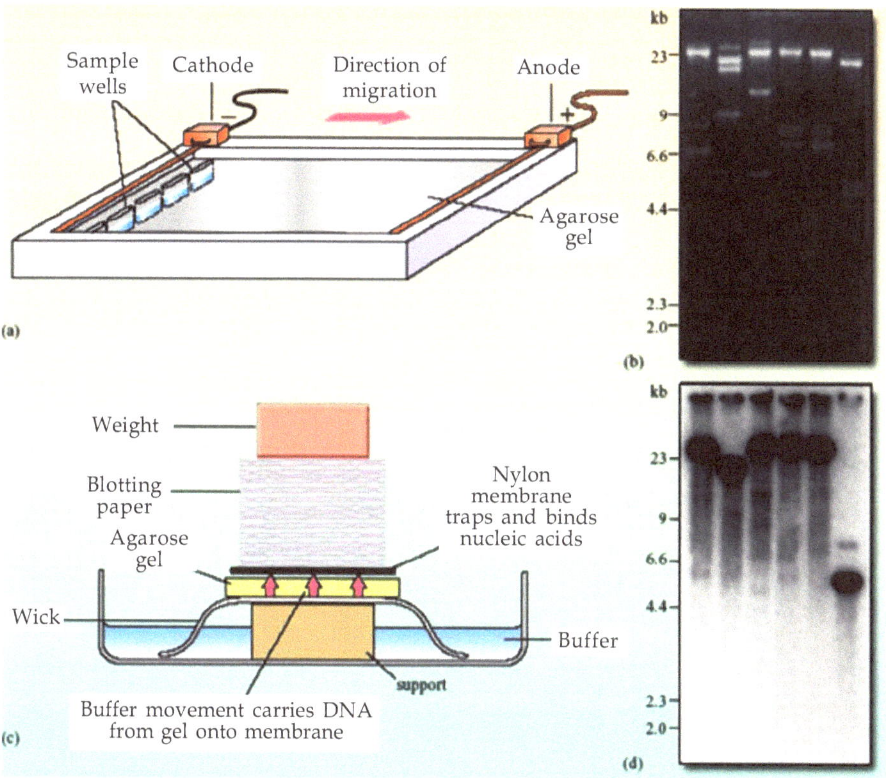

Fig.15: A DNA agarose gel electrophoresis system. The bands can be viewed by staining with ethidium bromide or can be blotted on nitrocellulose paper

- Rinse the gel briefly with dH_2O to remove any residual staining solution.
- Place the gel on a UV transilluminator for nucleic acid visualization and analysis. DNA/Ethidium Bromide complexes

may be illuminated with UV light of 254, 302, or 366 nm. Sensitivity decreases with illumination at higher wavelengths. However, nicking of DNA will increase below 302 nm.

Determination of DNA Concentration

To determine the concentration of DNA read the OD at 260nm in a spectrophotometer. An OD 260 of 1.0 give ~50 μg of dsDNA or ~37 μg of ssDNA. Check for purity of DNA by reading the OD 280 of the DNA solution as well. OD 260/OD 280 should give a value of 1.8 for pure DNA. Other values indicate the presence of contaminant such as proteins, RNA, and phenol.

Chapter 32

Extraction of DNA from Gel

There are many methods for extracting DNA from gel apart from using commercial kits, some methods however yield cleaner DNA than others.

1. Freeze squeeze

Excise band and place pieces of it in a home-made spin-column (a piece of filter placed in a PCR 0.5ml eppendorf, pierced the bottom of the PCR tube with a needle). Place the PCR tube at -20 °C for 15-20 mins or freeze in liquid N2. Then place the PCR tube into the 1.5 ml eppendorf and spin at top speed in a microfuge for 15 mins, collect DNA solution which can be EtOH ppt or use directly.

2. Powdered silica

Melt excised bands at 70°C cool and add equal volume of TE-buffered phenol and mix. Spin for 5 min, remove aqueous phase, repeat and EtOh ppt. the DNA.

3. Electroelution

Cut a small trough just ahead of the DNA band, place cellulose paper, or dialysis tubing, or affinity membrane, or just buffer containing 0.3 M NaOAc and 10% sucrose into the trough and run the gel at 150 volt for a minute or two until all the DNA has gone into the trough. Application of current will cause the DNA to migrate out of the agarose, but it will be trapped within the bag. When the DNA is out of the agarose, the flow of current is reversed for a few seconds to knock the DNA off of the side of the tubing. The buffer containing the DNA is then collected and the DNA precipitated with ethanol. Progress can be monitored using a transilluminator.

Chapter 33

Electroelution of DNA Gel into Dialysis Tubing

Principle

To retrieve and purify any specific DNA fragment from an agarose gel slice; the expected yield is from 50-75% of the amount in the gel slice.

Time required

1. Restriction digest - minimum of 2 hours
2. Electrophoresis - approx. 4 hours
3. Elution - 2-5 hours depending on fragment size
4. DNA Purification - 2 hours

Special Equiptment

Dialysis tubing (10mm x 50 ft.,) and clips

Electroelution buffer

- ❐ 1 M Tris, pH 7.5 12.0 ml
- ❐ 0.5 M EDTA 0.24 ml

- 1 M NaCl 3.0 ml
- qs to 600 ml dH_2O

Acetate cushion

- 3 M Na Acetate pH 4.8 480 ul
- 0.1 % Bromphenol Blue 40 ul

Electroelution of agarose fragments

1. Place gel slices in trough
2. Remove all air bubbles, then layer 80 μl of acetate cushion
3. Electroelute at: 120V for ~1Kb to 140V for >2.5Kb for 40 min for ~1Kb to 60 min for >2.5Kb
4. Collect ~300 ul of salt cushion, add 3X volumes of 95% ethanol to precipitate
5. Remove gel slices
6. Clean wells
7. Run for 10 min longer
8. Clean wells again
9. Rinse thoroughtly to remove any extraneous DNA

The restriction digestion

Use 5-10 μg DNA with the appropriate enzyme digest for a minimum of 2 hours, maximum of overnight. If available, use a restriction enzyme map of the DNA to be digested to determine what enzyme gives the best fragment separation and what the size of the fragments of interest should be. For example, cutting a plasmid into two similar size fragments makes separating the fragments difficult. The restriction maps can identifyanother enzyme which only cuts one of these fragments (usually the vector). By cutting with both enzymes, the unwanted fragment can be cut into smaller pieces giving better isolation of the wanted fragment.

Note: It is always helpful to run a small aliquot of the digest (250-500 ng) on a mini-gel to visualize the fragment pattern and determine how long to run the gel to get the desired separation, as well as to test for complete digestion.

Electrophoresis

- ❒ Generally, a 0.8 % TA gel will give sufficient separation for most digested fragments. Increase agarose to 1.2 % to isolate small fragments of 100 to 500 base pairs, decrease agarose to 0.6 % for fragments larger than 8 kb. 10 µg of digested DNA will need to be spread out over approximately. 6 cm of wells. Some combs have 6 cm "slots" or wells or this can be achieved by taping up 8-10 individual wells. Individual wells can be used but some DNA will be lost due to the trailing up effect from the well edges.
- ❒ Run the gel at 60-80 volts until desired separation has occurred (depending on fragment sizes). Stain the gel for 30 minutes with ethidium bromide, but do not photograph gel yet because short wave UV can damage the DNA.
- ❒ Place the gel on the transilluminator. The fragments should become visible.
- ❒ Use a clean razor blade to cut above and below each fragment of interest minimizing the amount of agarose in the slice. Then free the slice by cutting the ends. Leave a small amount of the fragment ends in the gel to verify fragment sizes in the photograph. Minimize the amount of time the DNA is exposed to UV by having sterile labeled tubes ready for the slices.
- ❒ Photograph the gel after the fragment slices have been removed. Sometimes the fragment bands are too faint to be seen with long-wave and you must use short-wave. Again, try to minimize any UV exposure.

Elution

- ❒ Cut a piece of dialysis tubing approximately 3 cm longer than the gel slice and clip one end. Gently push gel slice into open end and down to the clip. Add 300-500 ml of buffer (same as the gel, e.g., if the gel is 1 X TA, then use 1 X TA Buffer) so that the gel slice is completely immersed and there are no bubbles. Clip the open end

  Note: gel slices will sit better in the electrophoresis box if clips are positioned identically.
- ❒ Place gel slices parallel to the electrodes and fill the electrophoresis box with buffer (again, same as original gel) until all tubing is submerged. Then remove some buffer until clip edges stabilize and rest on bottom. Electroelute at 80-100 volts for 2-5 hours,

longer for large fragments. Monitor the movement of the DNA with a hand held UV source (long-wavelength).

- DNA may stick to dialysis tubing after current flow. Either reverse the electrodes or run for 15 seconds to dislodge the DNA or be careful to resuspend the DNA before removing the buffer. To resuspend DNA inside the tubing, first open one clip and carefully remove the agarose slice (which can be restained to verify elution) without losing any buffer. Then replace the clip and with your fingers, press along the length of tubing to mix the DNA with the buffer. Open the clip and remove the buffer with a pipetman. If low yields are suspected, add another 100 µl 1 X buffer to the empty tubing and repeat these steps to rinse out the tubing. Final volume should be less than 600 µl. If the agarose slice is much longer than 4 cm, the amount of buffer retrieved may be much more than 600 µl. In this case split the DNA buffer into two or three tubes so that no tube has more than 600 µl.

DNA cleaning and precipitation

- Add 1 volume (500 µl) of phenol and mix (NOTE: hard vortexing will shear large fragments). Spin in microcentrifuge at 14,000 rpm (at 4 degrees C or at room temperature) for five minutes. Phenol sinks and the DNA (in the aqueous phase) will be on top. Remove the top aqueous phase without pulling any debris at the interphase (don't try to get it all - the DNA quality will be better). Place the aqueous phase in clean eppendorf tubes.

- Add 1 volume (500 µl) of chloroform, mix, and spin in microcentrifuge and 14,000 rpm for two minutes. Chloroform will sink and the DNA will be in the aqueous phase. Remove the top aqueous phase leaving behind any debris at the interphase and transfer to clean eppendorf tubes.

- Add 1/10 volume (50 µl) 3M Na Acetate and 800 µl 95% EtOH (EtOH must be at -20 degrees C, or place tubes in the -20 degrees C freezer for 15 minutes after adding EtOH). Spin in microcentrifuge at 14,000 rpm, 4 degrees C for 30 minutes. Decant supernatant and add 500 µl 70% EtOH (must be cold!) to the pellet. Spin again for five minutes, decant the ethanol wash and invert the tubes to dry. For low yields or barely visible pellets, speed vacuum drying is recommended.

- When the DNA pellet is dry, add 20-50 μl 1 X TE; the pellets should resuspend fairly easily. Quantitate on a mini-gel with several lambda DNA standards and 1kb ladder for sizing.

Safety Precautions

Ethidium bromide is a powerful mutagen and moderately toxic. Gloves should be worn when handling gel and gel slices. UV can cause severe burns. Always wear eye protection. Phenol can cause severe burns and gloves should be worn.

Chapter 34

RNA Gel Electrophoresis

Principle

RNA is negatively charged at neutral pH and when an electric field is applied, it migrates towards the anode. RNA retains much of its secondary structure during electrophoresis unless it is first denatured. The addition of formaldehyde to the agarose gel maintains the RNA in its linear (denatured) form. The following protocol is based on Formaldehyde Agarose Gel Method.

Reagents required

- 20 x MOPS buffer 400 mM 3-(N-morpholino)
- Propanesulphonic acid,
- 160 mM sodium acetate,
- 20 mM EDTA, pH 7.0. Sterilise by autoclaving
- Ethidium bromide 10 mg / ml in sterile water
- Agarose

- Formaldehyde
- RNA loading buffer 1 x MOPS buffer (ie, 20 mM MOPS, 8 mM sodium acetate and 1 mM EDTA, pH 7.0), 7% (w/v) formaldehyde pH 4.0, 5% (v/v)
- Sterile glycerol, 50% (v/v)
- Deionised formamide
- 0.025% (v/v) of saturated aqueous bromophenol blue solution
- RNA samples

Equipment

- Gel caster/former
- Gel combs
- Electrophoresis power supply
- Micropipettes

Preparation of agarose gel

- Clean the gel-former and comb with distilled water, then 70% ethanol. Seal the edges of the gel-former with tape.
- Check that the comb sits approximately 1 mm above the gel-former when in situ 2 Make up enough buffer for both the electrophoresis tank and the gel to avoid any differences in ionic strength between them 3 For 100 ml of a 1.3% gel :
- Melt 1.3 g of agarose in 50 ml water.
- Add to 20 ml of formaldehyde 6, 5 ml of 20 x MOPS and 5 ml of ethidium bromide solution in a measuring cylinder.
- Make the volume up to 100 ml with Milli Q water 4 Swirl to mix and then pour the gel carefully, checking for air bubbles under or between the teeth7.
- The final gel should be between 3 mm and 5 mm thick.
- Allow it to set at room temperature for 30 - 45 minutes 8, 9, 5 When fully set, pour enough buffer to cover the gel surface by ~ 1 mm and allow to stand for a couple of minutes 10.
- Remove the comb carefully to avoid tearing the bottom of the wells (and subsequently losing the sample), and place the gel in the electrophoresis tank. Cover to a depth of ~ 1 mm

Preparation of RNA samples

5 ml of aqueous RNA is added to 15 ml of RNA loading buffer. The loading buffer can be diluted by up to 2 fold. Denature the RNA samples in loading buffer by heating to 70°C for 5 minutes and Chill on ice before loading the gel.

Preparation of RNA sample for electrophoresis

- ❒ Mix 1 volume of the 2X RNA Loading Dye Solution and 1 volume of RNA sample.
- ❒ Heat at 70°C for 10min.
- ❒ Chill on ice for 3 minutes and load onto the gel.

Applied voltage

At low voltage, the rate of migration of RNA is proportional to the voltage applied. As the voltage is increased, the mobility of larger molecules increases differentially. The effective range of separation therefore decreases with increasing voltage. For maximum resolution, run agarose gels at no more than 5V / cm (measured between the electrodes, not the gel length).

Denaturing formaldehyde gels in MOPS buffer

Freshly prepare 10X MOPS buffer:

- ❒ 0.4M MOPS (pH 7.0),
- ❒ 0.1M Sodium acetate,
- ❒ 0.01M EDTA (pH 8.0).

Note: 10 x MOPS Buffer is intended for use at 1x as the electrophoresis and gel running buffer during the separation of RNA on denaturing formaldehyde/agarose gels. 10x MOPS Buffer is composed of 200 mM MOPS, pH 7.0, 80 mM Sodium Acetate, and 10 mM EDTA, pH 8.0 in Molecular Biology Grade Water. For agarose gel electrophoresis, 10x MOPS Buffer should be diluted to a working concentration of 1x. For the preparation of an agarose gel, melt agarose in water, and add 10x MOPS Buffer and 12.3 M (37%) formaldehyde using a fume hood to respective final concentrations of 1x and 2.2 M.

Sample loading buffer

- ❒ Deionized formamide - 750 ml
- ❒ 10 x MOPS-Acetate - 150 ml

- Formaldehyde AR grade (37%) - 240 ml
- Sterile ddH_2O (DEPC treated & autoclaved) - 75 ml
- Glycerol (autoclaved) - 100 ml
- 3% Bromophenol blue (sterile) - 25 ml

Preparation of 1.2% agarose gel for RNA electrophoresis

- Sterile ddH_2O - 84.9 ml
- Agarose - 1.2 g,

Boil to dissolve and then add

- 10 X MOPs Acetate - 10.0 ml
- 37% Formaldehyde - 5.1 ml (0.66 M)
- Sterile ddH_2O to - 100.0 ml

Prepare 1% Agarose gel as follows

- stir 1g of agarose powder in 72ml of deionized water
- melt the agarose, and then add 10ml of 10X MOPS buffer and mix
- when the agarose solution cools to 60°C, add 18ml of fresh formaldehyde (37%) in a fume hood and mix thoroughly
- Pour the gel.
- Place the gel into an electrophoresis apparatus containing 1X MOPS buffer.

Heat the RNA samples and ladder at 70°C for 10min, and then chill on ice for 3 minutes. Load onto the gel.

Preparation of RNA sample for loading

- RNA (5 to 10 mg) - 5.0 ml
- Sample loading buffer - 25 ml

Denaturing glyoxal/DMSO gels in sodium phosphate buffer

- Prepare thick 1.0% Agarose gel in 0.01M sodium phosphate buffer, pH 7.0.
- Place the gel into an electrophoresis apparatus containing 0.01M sodium phosphate buffer, pH 7.0.
- Prepare for loading 25µl aliquots of the ladder/samples by adding:

- Glyoxal (40% solution) 4.5µl
- DMSO 12.5µl
- 0.1M sodium phosphate buffer, pH 7.0 2.5µl
- Mix and add RNA 3µl
- 2X RNA Loading Dye Solution 1µl
- DEPC-treated Water to 25µl

Electrophoresis of RNA sample

Load the RNA samples carefully into the slots. Connect the electrophoresis tank to a constant voltage power supply - RNA will run from black to red i.e. cathode to anode: make sure leads are on the right way round. Run at 1 - 5V / cm (measured as the distance between the electrodes) until the BPB dye front has migrated the appropriate distance. Check after the BPB has run 50% of the way down the gel. The ethidium bromide will migrate the opposite way to the RNA and long electrophoresis will remove much of the ethidium from the gel.

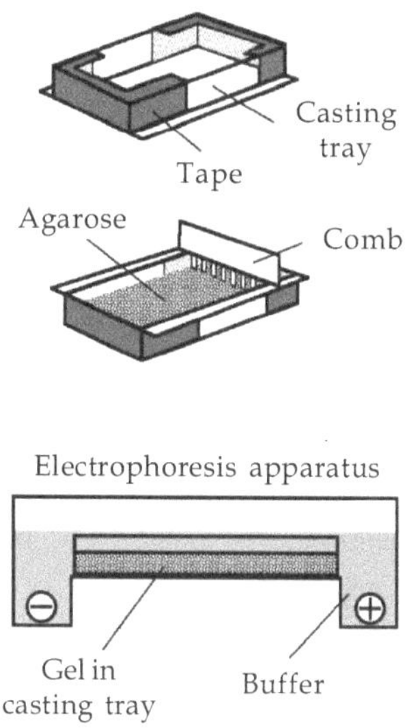

Fig.16: Components of an Electrophoresis apparatus

Incubate for 1 hour at 50°C and then cool down to room temperature.

- Load the samples on a gel.
- Run electrophoresis at 5V/cm until the bromophenol blue runs approximately two-thirds of the way down the gel

- Stain the gel in ethidium bromide solution (final concentration 0.5μg/ml) in 0.5M ammonium acetate for 15-30min
- Wash the gel in fresh 0.5M ammonium acetate solution for 15-30min.

Chapter 35

Polyacrylamide Gel Electrophoresis (PAGE)

Principle

Monomeric acrylamide (which is neurotoxic) is polymerised in the presence of free radicals to form polyacrylamide. The free radicles are provided by ammonium persulphate and stabilised by TEMED (N'N'N'N'-tetramethylethylene-diamine). The chains of polyacrylamide are cross-linked by the addition of methylenebisacrylamide to form a gel whose porosity is determined by the length of chains and the degree of crosslinking. The chain length is proportional to the acrylamide concentration: usually between 3.5 and 20%. Cross-linking bisacrylamide is usually added at a ratio of 2g BIS 38g acrylamide.

Two types of polyacrylamide gel in general use:

- *Non-denaturing gels*

These are run at low voltages - 8V/cm - and 1 x TBE to prevent denaturation of small fragments of DNA by the heat generated in the gel during electrophoresis. The rate of migration is approximately inversely proportional to log10 of their size. However, the base sequence composition can alter the electrophoretic mobility of DNAs such that two DNAs of the same size may show up to a 10% difference in electrophoretic mobility

- *Denaturing gels*

These gels are polymerised with a denaturant that suppresses base pairing in nucleic acids - this is usually urea but can be formamide. Denatured DNA migrates through the gel at a rate which is almost completely independent of its composition or sequence. These gels are used for the analysis of sequencing reactions, DNAase protection assays and purification of radiolabelled DNA and DNA probes.

Advantages over agarose gels

1. Resolving power is such that DNA molecules differing in length by 1 base in 500 ie 0.2% can be effectively separated.
2. They can hold up to 10 mg of DNA per slot (20 mg of RNA) without loss of resolution, a much larger amount than agarose gels

The recovered DNA or DNA is extremely pure.

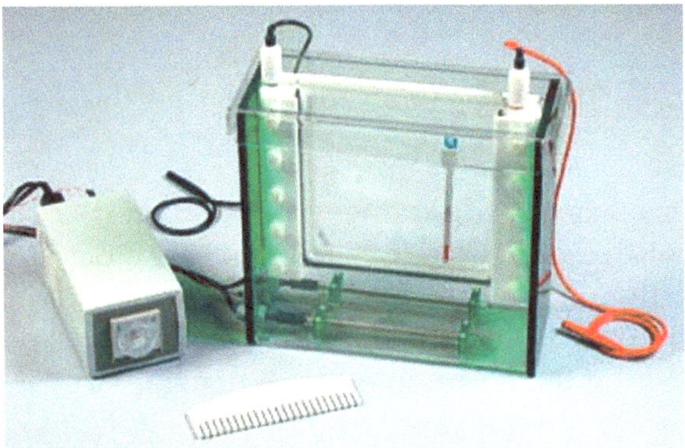

Fig.17: Vertical SDS PAGE Electrophoresis Unit

Table 10 : % of acrylamide and separation range

% of acrylamide (w/v) with BIS at 1:20	Effective range of separation - bp
3.5	1 000 - 2 000
5.0	80 - 500
8.0	60 - 400
12.0	40 - 200
15.0	25 - 150
20.0	6 - 100

Preparation of solutions

Acrylamide-bisacrylamide solution

Acrylamide	-	29.2 g
Bis acrylamide	-	00.8 g
dH_2O to	-	100 ml

Stir the solution for 15 min with a small amount of activated charcoal and later filter through nitrocellulose membrane filter, stored at 4° C in a brown bottle. (Never store the solution for a long time, as acrylamide becomes acrylic acid).

4x separating gel buffer (1.5 M Tris, pH 8.8)

Dissolve 18.15 g of 1.5 M Tris in 80 ml dH_2O, adjust the pH to 8.8 with conc. HCl and make up the volume to 100 ml.

4x stacking gel buffer (0.5 M Tris, pH 6.8)

Dissolve 3.0 g of 0.5M Tris in 30 ml dH_2O, adjust the pH to 6.8 with conc. HCl and make up the volume to 50 ml.

Initiator: (10% APS)

Ammonium persulphate	-	0.1 g
dH_2O	-	1.0 ml

Dissolve 0.1g of APS in 1.0ml of dH_2O and store in ice.

Catalyst : TEMED is supplied in brown bottle and stored at 4° C.

Separating gel overlaying solution

n - or iso-butanol	-	50 ml
dH_2O	-	50 ml

Mix 50ml of n - or iso-butanol with 50ml of dH2O. Separate the water saturated butanol (upper layer) after shaking the mixture vigorously and then allowing it to settle.

Bromophenol blue (0.1%)

Dissolve 5 mg bromophenol blue in 5 ml water.

2 x sample buffer

- Tris (0.5 M, pH 6.8) - 2.5 ml (4 x Stacking gel buffer)
- SDS (10%) - 4.0 ml

- Glycerol (100%) - 2.0 ml
- β-mercaptoethanol - 0.8 ml (or IM DDT – 0.5 ml)
- Bromophenol blue (0.1%) - 300 ìl
- dH_2O (400 μl) to - 10.0 ml

Tank buffer

- Tris - 6.05 g
- Glycine - 28.80 g
- 10% SDS - 10.0 ml or (1.0 g)
- dH_2O - 1000 ml

Staining solution

- Coomassie blue (R-250) - 0.3 g
- Methanol (AR) - 80 ml
- Glacial acetic acid - 20 ml
- dH_2O - 100 ml

Destaining solution

- Acedic acid - 100 ml
- Methanol - 300 ml
- dH_2O - 1.0 litre

Gel storing solution

- Acetic acid - 15 ml
- dH_2O - 200 ml

Standard proteins (2 μg/μ1)

- Phosphorylase a (97.4 kDa) - 2 mg/ml
- BSA (66 kDa) - 2 mg/ml
- Egg albumin (45 kDa) - 2 mg/ml
- Carbonic anhydrase (29 kDa) - 2 mg/ml
- Lysozyme (14.4 kDa) - 2 mg/ml

Preparation of Separating gel – 10% (10 ml is required for baby gels)

- dH_2O - 4.00 ml
- 30% acrylamide:bisacrylamide - 3.33 ml

- 4 x Tris (pH 8.8) - 2.50 ml
- TEMED - 5 µ1
- 10% SDS - 100 µ1
- 10% APS (100 mg/ml) - 50 µ1

10.00 ml

Preparation of Stacking gel – 4% (5 ml is required for baby gels)

- dH_2O - 3.00 ml
- 30% acrylamide: bis - 0.67 ml
- 4 x Tris (pH 6.8) - 1.25 ml
- TEMED - 2.5 µ1
- 10% SDS - 25 µ1
- 10% APS - 50 µ1

The method of pouring and running a neutral polyacrylamide gel

- Clean the glass plates and spacers thoroughly. Hold the plates by the edges or wear gloves, so that oils from the hands do not become deposited on the working surfaces of the plates. Rinse the plates with deionized water and ethanol and set them aside to dry.

  Note: The glass plates must be free of grease spots to prevent air bubbles from forming in the gel.
- Assemble the glass plates with spacers in gel caster.
- Prepare the gel solution with the desired polyacrylamide percentage according to the table below, which gives the amount of each component required to make 12 ml (sufficient for 2 Hoefer minigels of 1mm thickness):

Table: 11 Volume of Reagents Used to Cast Polyacrylamide Gels:

Gel %	30% Acylamide (29:1)	H_2O (ml)	5x TBE (ml)	10% APS (µl)	TEMED (µl)
8%	3.2ml	6.4	2.4	200	10
10%	4.0ml	5.6	2.4	200	10
12%	4.8ml	4.8	2.4	200	10

- Wear gloves. Work quickly after addition of TEMED to complete the gel before the acrylamide polymerizes.

- Immediately insert the appropriate comb into the gel, being careful not to allow air bubbles to become trapped under the teeth. The tops of the teeth should be slightly higher than the top of the glass. Clamp the comb in place with strong paper clips. If necessary, use the remaining acrylamide gel solution to fill the gel mold completely. Make sure that no acrylamide solution is leaking from the gel mold.
- Let the acrylamide to polymerize for 30-60 minutes at room temperature.
- After polymerization is complete, surround the comb and the top of the gel with paper towels that have been soaked in 1x TBE. Then seal the entire gel in Saran Wrap or plastic bag and store it at 4°C until needed.

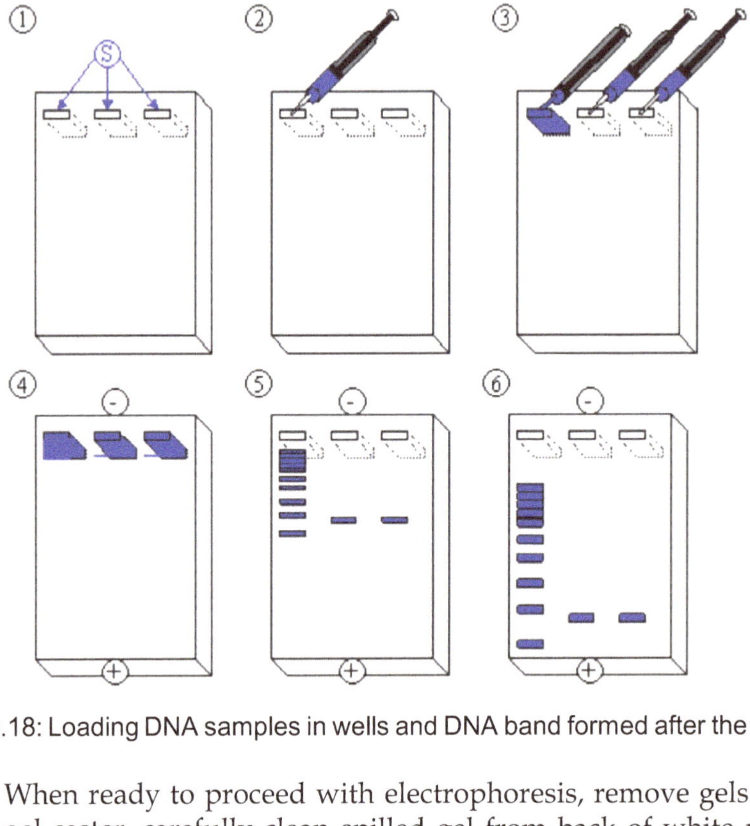

Fig.18: Loading DNA samples in wells and DNA band formed after the run

- When ready to proceed with electrophoresis, remove gels from gel caster, carefully clean spilled gel from back of white plates and insert gels into Hoefer gel box. Add running buffer and carefully pull the combs from the polymerized gel.

Note: It is important to use the same batch of electrophoresis buffer in both of the reservoirs and in the gel. Small differences in ionic strength or pH produce buffer fronts that can greatly distort the migration of DNA.

- ❒ Use a Pasteur pipette or a syringe to flush out the wells once more with 1x TBE. Mix the DNA samples with the appropriate amount of gel-loading buffer. Load the mixture into the wells using a micropipette equipped with a drawn-out plastic tip.

  Note: Avoid expelling the entire sample from the loading device, as this almost always produces air bubbles that blow the sample out of the well. It is important not to take too long to complete loading the gel; otherwise, the samples will diffuse from the wells.

- ❒ Connect the electrodes to a power pack, turn on the power, and begin the electrophoresis run. Nondenaturing polyacrylamide gels are usually run at voltages between 1 V/cmand 8 V/cm. If electrophoresis is carried out at a higher voltage, differential heating in the center of the gel may cause bowing of the DNA bands or even melting of the strands of small DNA fragments.
- ❒ Run the gel until the marker dyes have migrated the desired distance. Turn off the electric power, disconnect the leads, and discard the electrophoresis buffer from the reservoirs.
- ❒ Detach the glass plates. Lay the glass plates on the bench. Use a spacer or plastic wedge to lift a corner of the upper glass plate. Check that the gel remains attached to the lower (white) plate. Pull the upper plate smoothly away. Remove the spacers.
- ❒ Stain gels with EtBr.

Chapter 36

Polyacrylamide Gel Electrophoresis of RNA

Denaturing polyacrylamide/urea gels in TBE buffer

Prepare 20ml of a 5% polyacrylamide gel containing 7M urea by adding:

- 47.5% acrylamide:
- 2.5% bis-acrylamide solution 2ml
- 10M urea 14ml
- 10X TBE buffer 2ml
- 10% freshly prepared ammonium persulfate 0.2ml
- deionized water 1.8ml

Mix and add 10µl TEMED

- Mix again and pour the gel carefully avoiding the formation of air bubbles.
- Insert the comb into the acrylamide and allow the gel to polymerize for at least 1 hour.
- Fill the electrophoresis apparatus with 1X TBE buffer

- Heat the RNA samples (premixed with equal volume of 2X RNA Loading Dye Solution) at 70°C for 10min, and chill on ice for 3 minutes. Load onto the gel.
- Run electrophoresis at 8V/cm for about 1 hour.
- Soak the gel for about 15 minutes in 1X TBE to remove urea prior to staining.
- Stain the gel in 0.5μg/ml ethidium bromide in 1X TBE solution for 15 min.

Chapter 37

Silver Nitrate Staining

The principle of silver nitrate staining of proteins is based on the reduction of silver ions to its metallic form. (Silver ions in alkaline conditions bind with proteins through å-amino group of lysine and sulphur groups of cysteine and methionine residues). When the complexed silver ions are reduced in the presence of formaldehyde, they become metallic silver, which is seen on the gel as bands. Staining of proteins with silver nitrate after SDS-PAGE is quicker and ~ 10 times more sensitive than Coomassie brilliant blue staining and that is the reason it is widely used in research laboratories these days.

The procedure is as follows:

Preparation of solutions

1. Mix 50 ml water and 50 ml methanol (Fixing solution-1)
2. Mix 50 ml water and 35 ml methanol and 15 ml 37% formaldehyde (Fixing solution-2)
3. Sodium thiosulphate 0.22% (20 mg in 100 ml)- dilute 1:10 from 0.2% stock.

4. Dissolve 100 mg silver nitrate in 100 ml water (0.1% Staining solution)
5. Dissolve 3 g K_2CO_3 OR Na_2CO_3 in 100 ml water and add 2 ml of 0.02% thiosulphate solution and 0.05 ml 37% formaldehyde (freshly prepared) (Developing solution)
6. Dissolve 5 g citric acid in 100 ml water (Stopping solution)
7. Mix 10 ml glacial acetic acid 10 ml methanol and 80 ml water (Gel storing solution)
8. (For staining small amount of gels the volume of the above solutions can be reduced to half)

Procedure

1. Transfer the gel to the Fixing solution-1 after electrophoresis and incubate for 30 min. (The gel can be stored for a long time in this solution) followed by dipping the gel in Fixing solution-2 for 10 min.
2. To remove excess methanol and formaldehyde Wash the gel twice with ~50 ml water for 5 min each time
3. Transfer the gel in 0.02% thiosulphate solution for one min and rinse with water.
4. Immerse the gel in silver nitrate solution for 15 min and later rinse thoroughly with distilled water to remove excess silver nitrate.
5. Incubated the gel in 100 ml of developing solution consisting of 3% K_2CO_3 solution containing 2.0 ml of 0.2% thiosulphate solution 0.05 ml formaldehyde until you see the bands.
6. Stop the reaction immediately by adding 100 ml of the citric acid solution.
7. The gel can be stored in 10% acetic acid methanol solution. (For long term storage 10% glycerol can be included).

Chapter 38

DNA Sequencing

Principle

Determination of the precise sequence of nucleotides in a sample of DNA is called "DNA Sequencing". The most popular method for doing this is called the "Dideoxy method or Sanger Method" (named after Frederick Sanger, who was awarded the 1980 Nobel Prize in chemistry for inventing it). DNA is synthesized from four deoxynucleotide triphosphates. The top most figure is a normal nucleotide: deoxythymidine triphosphate (dTTP). During polymerization each new nucleotide is added to the 3' -OH group of the last nucleotide added.

The dideoxy method has been coined after the synthetic nucleotides that lack the -OH at the 3'carbon atom of the sugar and it plays an important role in this process atom (red arrow). In the figure below dideoxythymidine triphosphate – ddTTP, a dideoxynucleotide has been shown that can be added to the growing DNA strand. But the chain elongation ceases when the next nucleotide tries to get attached to the former as there is no 3' -OH for the next nucleotide to be attached to. Hence, the dideoxy method is also called the chain termination method.

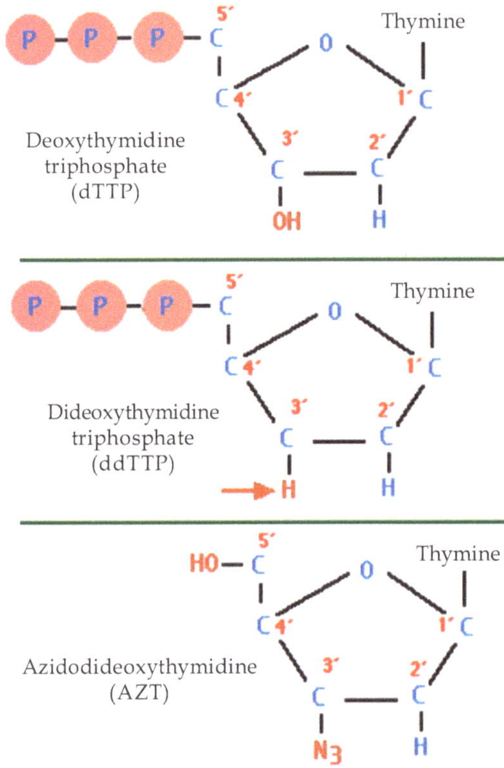

Chain elongation proceeds normally until because all four normal nucleotides are present in the solution. When, by chance, DNA polymerase inserts a dideoxy nucleotide instead of the normal deoxynucleotide, the chain termination occurs. If the ratio of normal nucleotide to the dideoxy versions is high enough, some DNA strands will succeed in adding several hundred nucleotides before insertion of the dideoxy version halts the process.

DNA fragments are separated by length at the end of the incubation period (from longest to shortest). The resolution is so good that a difference of one nucleotide is enough to separate that strand from the next shorter and next longer strand. Each of the four dideoxynucleotides fluoresces a different color when illuminated by a laser beam and an automatic scanner provides a printout of the sequence.

The procedure

DNA sequencing is typically performed using a kit. Consult the manufacturer's instructions for the precise protocol to use. The procedure will involve mixing an oligonucleotide primer with the template DNA and allowing a polymerase to synthesize a new strand of DNA. The procedure is thus a variation on the PCR method, with two modifications:

1) Only one oligonucleotide is necessary, and
2) Four separate reactions must be run,

Each reaction should contain a single dideoxynucleotide for specific chain termination (in addition to the four normal deoxynucleotides).

The DNA to be sequenced is prepared as a single strand.

Requirements

1. A mixture of all four normal (deoxy) nucleotides in sufficient quantities

- ❒ dATP
- ❒ dGTP
- ❒ dCTP
- ❒ dTTP

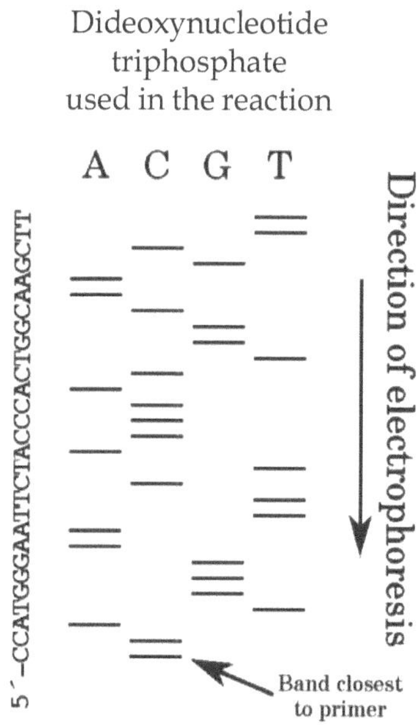

Fig.19: A representation of a sequencing gel autoradiograph is shown

2. A mixture of all four dideoxynucleotides, in limited quantities and each labeled with a "tag" that fluoresces a different color:
    - ddATP
    - ddGTP
    - ddCTP
    - ddTTP
3. DNA polymerase I

Section-6

Proteins

Chapter 39

SDS PAGE for Protein Separation

Principle

The SDS polyacrylamide gel electrophoresis (PAGE) performs:

- Separation of proteins according to their molecular weight
- Ascertain whether a given protein is actually present in a sample,
- Assessment of purity of the preparation
- Estimation of the approximate quantity of the protein, and
- Measurement of the size of the protein.

The separation of proteins usually involves the use of polyacrylamide as the matrix. Polymerization of acrylamide monomers in the presence of N,N´-methylene bis-acrylamide results in the formation of Polyacrylamide. The two double bonds of bis-acrylamide cross-link the polyacrylamide chains, resulting in the formation of a gel matrix rather than a simple linear polymer.

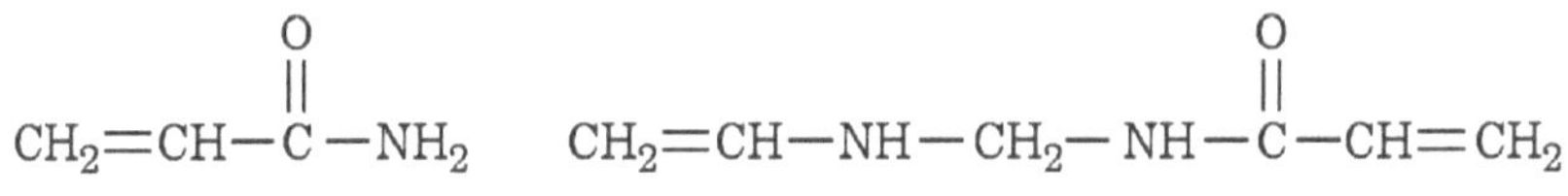

Acrylamide N, N′-methylene bisacrylamide

The polymerization reaction uses free radicals released by an unstable ammonium persulfate. The sulfate radicals formed then react with tetramethylethylenediamine (TEMED), forming TEMED radicals which react with acrylamide to start the polymerization process.

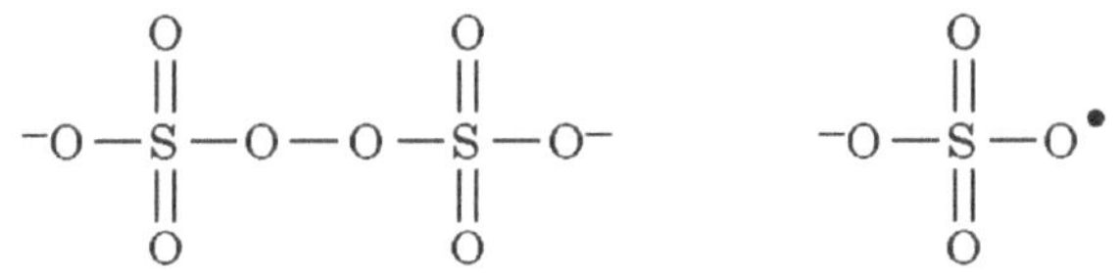

Ammonium persulfate Sulfate radical

The formation of the matrix depends upon the concentration of acrylamide monomer and bis-acrylamide cross-linker. Denser matrix results when the amounts of these chemicals are more. To separate smaller proteins one should use denser matrices, as larger proteins may find the pores too small to enter or pass, hence may fail to pass. The following table indicates ranges for different gel densities.

Table 12: Molecular weight and % acrylamide

Percent acrylamide	Useful Molecular weight range
7	30,000 to 200,000
10	20,000 to 150,000
12	10,000 to 100,000
15	5,000 to 70,000

Proteins have different proportions of charged side-chains; hence do not have constant charge-to-mass ratios, also have different three-dimensional shapes. In order to measure molecular weight, it is necessary to induce the formation of a similar shape and charge-to-mass ratio. Disruption of the three dimensional structure of the protein will take place if you boil it in the presence of the detergent sodium dodecyl sulfate (SDS) and the reducing agent mercaptoethanol (which reduces disulfide bonds). Treatment with SDS and mercaptoethanol will result in the formation of denatured protein monomers; it is these protein monomers that are separated on the SDS PAGE.

$$CH_3\text{–}CH_2\text{–}CH_2\text{–}CH_2\text{–}CH_2\text{–}CH_2\text{–}CH_2\text{–}CH_2\text{–}CH_2\text{–}CH_2\text{–}CH_2\text{–}CH_2\text{–}SO_4^- \;\; Na^+$$

Sodium Dodecylsulfate

Note: SDS molecule has a negative charge at the pH used for electrophoresis, the use of SDS results in a large negative charge that overshadows any intrinsic charge present in the protein.

Detection of proteins

Proteins can be detected at the end of electrophoresis by a dye called Coomassie Blue R-250. Staining can be done by immersing the gel in a solution of Coomassie blue in acetic acid and methanol. Acetic acid and methanol cross-link the proteins into the gel so that they do not diffuse. Once the protein is stained, the gel is destained in a solution of acetic acid and methanol that removes excess Coomassie blue.

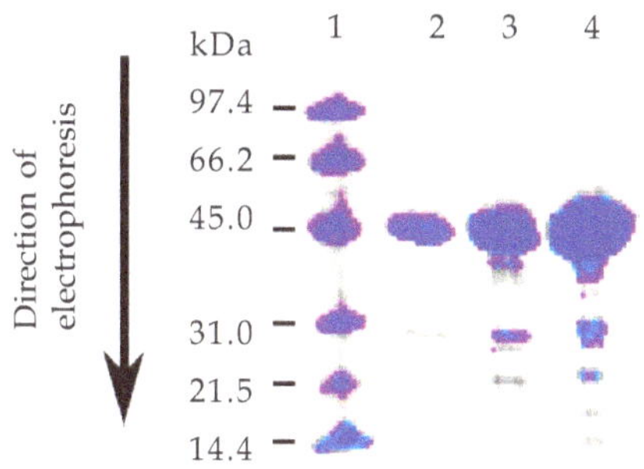

Fig.20: SDS gel with Lane 1 proteins of known molecular weight called molecular weight standards. Lanes 2, 3, and 4 contain increasing concentrations of an experimental protein sample.

Standard proteins of known molecular weights can be used to calibrate the migration of proteins of differing sizes on the gel. The migration is always inversely proportional to the log of the molecular weight. Hence one can plot a graph of log molecular weight *versus* distance travelled by the proteins of known size can act as a standard curve for measuring the molecular weights of unknown proteins. An example of an SDS PAGE standard curve is shown below:

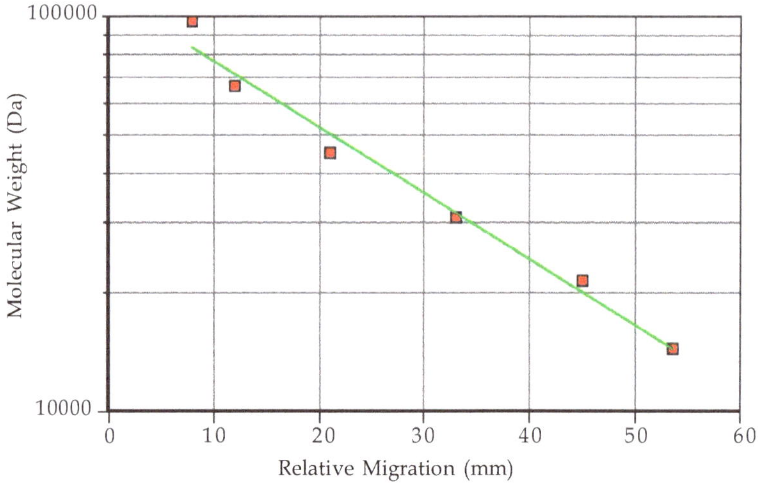

Fig. 21: Graph of log molecular weight *versus* migration distance of known size proteins

Note: The molecular weight of proteins determined will be that of the individual monomers as the proteins within an SDS polyacrylamide gel are denatured.

A commonly used set of molecular weight standards for SDS PAGE experiments is shown in the table below:

Table:13 Molecular weight of protein standards

Standard protein	Molecular weight (Da)
Rabbit muscle phosphorylase b	97400
Bovine serum albumin	66200
Chicken ovalbumin	45000
Bovine carbonic anhydrase	31000
Soybean trypsin inhibitor	21500
Chicken lysozyme	14400

The gel consists of two layers:

- Stacking gel: The top layer which concentrates all of the protein in a narrow region. The stacking gel also contains the wells into which the samples are loaded.
- Resolving gel: The bottom layer performs the actual separation of the proteins by molecular weight

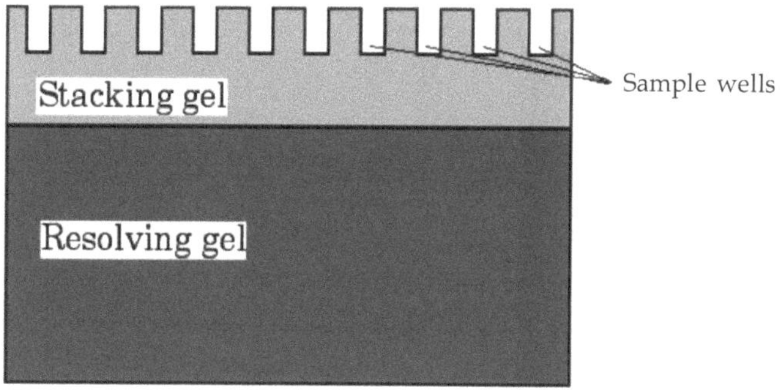

Fig. 22: A picture of an SDS PAGE gel

Chemistry of the stacking gel

It has low pH (6.8) and acrylamide percentage (6%). The neutral zwitterionic form of Glycine present in the electrophoresis tank buffer, at this low pH, is not an effective carrier of current. The highly charged chloride ions also present migrate briskly toward the anode. The SDS-coated protein molecules and the dye, which have charge-to-mass ratios greater than that of the glycine but less than that of Cl^{-1}, migrate behind the Cl^{-1} and ahead of the glycine. This causes proteins to concentrate as a thin band between the Cl^{-1} ions and the glycine molecules. Apart from this the low acrylamide percentage of the stacking gel does not inhibit the proteins from moving freely through the gel matrix. Therefore, as the sample reaches the end of the stacking gel it appears as a thin blue band.

Chemistry of the resolving gel

Resolving gel is at a higher pH (8.8) with higher concentration of acrylamide needed to separate of proteins of different sizes. When the stacked samples enter the resolving gel having higher pH, the negatively charged glycine molecules then migrate along with the Cl^{-1} ions. The protein samples lag behind and are separated by the sieving effect of the gel. The tracking dye (bromophenol blue) migrates faster than the proteins. Electrophoresis should be stopped when the dye reaches the end of the gel, o prevent the proteins from running off of the gel.

Gels have fixed protein loading. Therefore overloading should be avoided for smooth running of the sample. Similarly loading too little protein makes the protein bands difficult to detect. In choosing the amount of protein to load, you need to consider the number of proteins in your sample; if your sample contains many proteins, you can load more total protein.

Chemicals required for SDS PAGE

A. 10% Resolving gel

- 1.5 ml 1.5 M Tris-HCl, pH 8.8
- 2 ml 30% acrylamide
- 0.06 ml 10% SDS
- 2.44 ml water

B. Add last to initiate reaction

- 30 μl 10 % ammonium persulfate
- 5 μl TEMED

C. 5x Sample Buffer

- 60 mM Tris-HCl, pH 6.8
- 25% glycerol
- 2% SDS
- 14.4 mM -mercaptoethanol
- 1% bromophenol blue

D. Electrophoresis tank buffer

- 25 mM Tris
- 192 mM glycine, pH 8.8
- 0.1% SDS

E. Stacking gel

- 0.875 ml 1.0 M Tris-HCl, pH 6.8
- 0.583 ml 30% acrylamide
- 0.035 ml 10% SDS
- 2.007 ml water

F. Add last to initiate reaction

- 30 μl 10 % ammonium persulfate
- 5 μl TEMED

G. Coomassie Staining Solution

- 10% Acetic acid,
- 25% Methanol,
- 0.05% Coomassie R-250 or Bio-Safe Coomassie Blue solution

H. Instruments & Accessories

- Electrophoresis apparatus
- Gel loading tips
- Weigh boats for gel staining
- 100°C water bath
- Microcentrifuge tubes

Chapter 40

Cell Lysis for Protein Purification

Cell Lysis Procedure

Pellet the cells in a microfuge tube: add 1.5 ml culture, centrifuge for 15 seconds at maximum speed, discard the supernatant, and then repeat the procedure. This will result in the cells from 3 ml of culture in a single tube.

- Resuspend the cells in 1 ml Lysis buffer.
- Take 300 µl aliquot for Bradford protein assay.
- Add 50 µl lysozyme, and mix gently.
- After 10-20 minutes at room temperature, microfuge for 10 minutes to pellet cell debris.
- Use supernatant for LDH assay.

Required Materials

Lysis Buffer

- 50 mM Tris-HCl, pH 8.0
- 10 mM EDTA
- 2 mM Dithiothreitol

Lysozyme

- 50 mg/ml lysozyme in Lysis buffer
- LDH assay reagents
- Bradford assay reagents.
- Microfuge
- 1.5 ml microfuge tubes

Chapter 41

Western Blot Analysis of Proteins

Principle

Western blot analysis (also called protein immunoblot) is a useful tool for the detection of proteins in biological samples. It uses gel electrophoresis to separate native or denatured proteins by the length of the polypeptide (denaturing conditions) or by the 3-D structure of the protein (native/ non-denaturing conditions). The proteins are then transferred to a membrane (typically nitrocellulose), where they are probed (detected) using antibodies specific to the target protein. It is a technique for analyzing and identifying protein antigens: the proteins are separated by electrophoresis in polyacrylamide gel, then transferred ("blotted") onto a nitrocellulose membrane or treated paper, where they bind in the same pattern as they formed in the gel. The antigen is overlaid first with antibody, then with anti-immunoglobulin or protein A labeled with radioisotope, fluorescent dye, or enzyme.

The method originated from the laboratory of George Stark at Stanford. The name "Western blot" was given to the technique by W. Neal Burnette and is a play on the name "Southern blot", a technique for DNA detection developed earlier by Edwin Southern. Detection of RNA is termed northern blotting and the detection of post-translational modification of protein is termed "Eastern blotting".

The efficiency of protein transfer from the SDS-PAGE gel to the support membranes can be affected by the gel thickness and the total acrylamide concentration. Transfer buffers can also affect the efficiency of protein migration through the gel. While methanol is necessary to prevent gel swelling with heating, and to keep protein adsorbed to membrane, methanol-containing buffers often result in the decrease in pore size and may even precipitate proteins within the gel. Poor transfer can be remedied by using methanol-free transfer buffer.

In western blot, the transfer of proteins in membranes is mostly followed by immunoassay step where specific antibody is used to probe for specific antigens. Immunoassay involves the following steps:

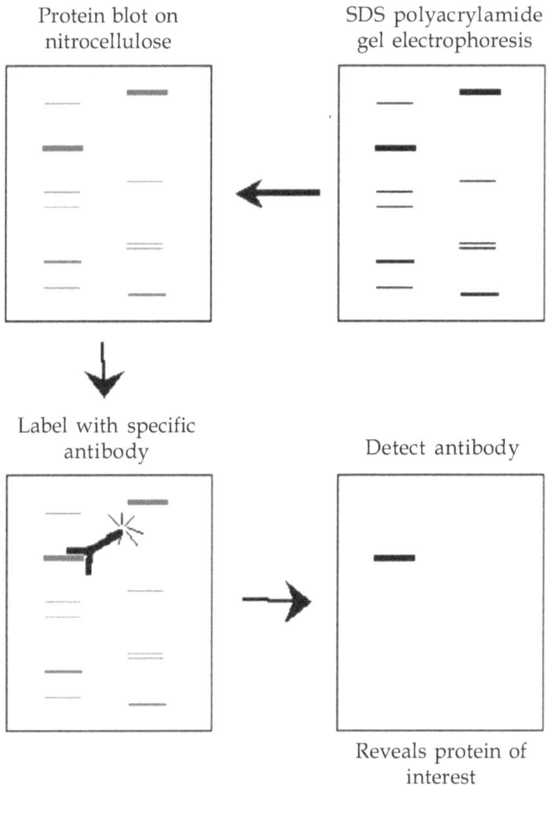

Fig. 23: Steps in Western Blotting

1. Blocking

The transfer membranes are blocked with a concentrated protein solution (e.g. ovalbumin, BSA, haemoglobin, non-fat milk powder) to prevent further non-specific binding of proteins [non-protein (polyvinylpyrrolidone) in the presence or absence of non-ionic detergent (such as Tween 20), may also be used for blocking]. This step is followed by

2. Incubation

Incubation of the membrane in a diluted antibody solution, washing of the membrane, incubation in diluted conjugated probe antibody or other detecting reagent, further washing, and finally detection.

Basic assay summary

- Proteins are separated by gel electrophoresis
- Proteins are transferred to a membrane, spaced out by the electrophoresis
- The blot is incubated with non fat dry milk (NFDM) to bind to any remaining sites on the membrane. An antibody is then added to the solution to bind to its specific target protein. The antibody has an enzyme (e.g. alkaline phosphatase or horseradish peroxidase) or dye attached to it that serves to aid in visualization.
- The location of the antibody is revealed by incubating it with a substrate and exposing a film to the reaction
- The band of interest is seen by developing the film
- Band density is compared using specialized software and a good quality optical scanner.

Note: Depending on the antobody application protocol one applies (i.e. overnight incubation or 1 hour antibody incubation with the membrane), the protocol takes 1 or 2 days to complete.

Procedure for running the SDS PAGE and blotting

Chemicals required

1. Transfer Buffer

- 192 mM Glycine
- 25 mM Tris pH 8.3

- 1.3 mM SDS
- 20% methanol

2. Blocking solution

- 1% gelatin or 3% BSA in TBST
- 0.02% sodium azide)

3. Tris Buffered Saline (TBS)

- 50 mM Tris-HCl, pH 7.5
- 150 mM NaCl

4. TBST

TBS with 0.1% Tween-20

5. 5x Sample buffer

6. Coomassie Staining Solution

7. 10% SDS polyacrylamide gels

8. Electrophoresis tank buffer

9. Hardware required

- Electrophoresis apparatus
- 100°C Water bath
- Gel loading tips
- Weigh boats for gel staining
- Hoefer Semi-Dry Transfer apparatus
- Filter paper
- Transfer membrane

Process

Mark the location of wells before adding tank buffer. After adding the tank buffer, remove the bubbles from the bottom of the gel.

1. Prepare samples

In most cases, it is necessary to run at least one SDS-PAGE gel to determine the appropriate concentration of protein to load for the Western blot. Ideally, Western blotting involves concentrations of protein that yield sharp, easily detectable, individual bands. Once the correct protein concentration is determined, enough of each sample should be prepared to allow two lanes of each sample to be run.

2. Loading samples

A common method of running a Western blot is to cut the SDS PAGE in half after electrophoresis, with one half stained with Coomassie blue and the other transferred for Western blotting. Therefore, load to gel to yield two identical halves.

3. Run the gel

Attach the electrodes to the gel apparatus, and turn on the power supply. Run the gel at constant current (15 mA per gel) until the tracking dye reaches the end of the gel. Be careful: the high current levels used during gel electrophoresis can be dangerous!

4. Disassemble the gel

Turn off the power supply and remove the electrodes. Remove the gel sandwich. Remove one spacer, and insert it in one corner between the glass plates. Use the spacer to gently pry off one glass plate. Remove the stacking gel and discard it.

5. Separate the gel

Cut the resolving gel in half. Place half of the gel (the half with the unstained protein standards) in a weigh boat with Coomassie blue stain and incubate overnight. The other half of the gel (the one with the stained protein markers) will be transferred to the membrane.

Transfer the protein

Chemicals required

1. Tris Buffered Saline (TBS)

- 50 mM Tris-HCl, pH 7.5
- 150 mM NaCl

2. TBST
- TBS with 0.1% Tween-20

3. Antibody Buffer
- 1% gelatin in TBST

4. Primary Antibody

5. Secondary Antibody
- Biotin-conjugated Goat anti-rabbit immunoglobin 10 μl/30 ml antibody buffer

6. Alkaline phosphatase
- AP, 10 μl/30 ml antibody buffer

7. Development Solution
- 200 μl BCIP (15 mg/ml in DMF)
- 200 μl 30 mg/ml NBT in 70% DMF
- 20 ml 100 mM Tris-HCl,
- 200 mM NaCl,
- 5 mM MgCl2, pH 8.0

Hardware required
- Shaker
- Weigh boats

Procedure
- Remove SDS from the gel by washing the gel with urea buffer for 30 min.
- Measure the size of the gel and cut 6 sheets of Whatman No.3 paper and 2 sheets of nitrocellulose membrane.
- Assemble the sandwich as given in fig.:
- Put rubber bands to keep the sandwich intact and secured.
- Keep the sandwich in transfer buffer kept in a tray.
- Keep a lightweight on top of the sandwich to prevent it from floating and to ascertain that it is kept immersed completely.

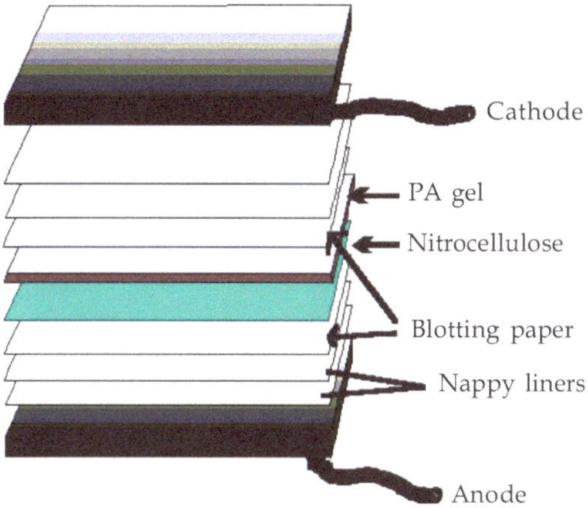

Fig. 24: Assembly of sandwich in SDS PAGE

Avoid trapping any air bubble while preparing the sandwich assembly.

- Keep for transfer for 12 to 18 hrs.
- Remove nitrocellulose membrane and proceed for Ab labelling or can be stored in a box for more than 6 months.

Electrophoretic Transfer

- Soak the gel, Whatman paper and nitrocellulose (NC) transfer buffer (80% running buffer + 20% methanol)
- Transfer protein electrophoretically using the transfer apparatus.

Procedure for developing western blot

1. Take the NC paper and shake in the blocking agent for 1 - 2 hrs. (3% milk in TBS, 10ml Tris-HCl, and pH 7.3, contains 0.15M NaCl).
2. Transfer this paper to 1% primary Ab solution made in blocking agent. Keep at room temperature for 1 hr. with constant shaking.
3. Rinse the paper in distilled water and keep in another container with TBS for 10min. at room temperature.
4. Repeat step 3 with fresh TBS.
5. Add secondary Ab conjugated with alkaline phosphatase (1:2000 dilutions) in 1% milk in TBS.

6. Shake for 1hr. at room temperature.
7. Rinse in distilled water and wash in TBS 10 min. twice.
8. Add the substrate solution (NBT + BCIP) in dark.
9. Colour develops within ten min. Take a photograph immediately.

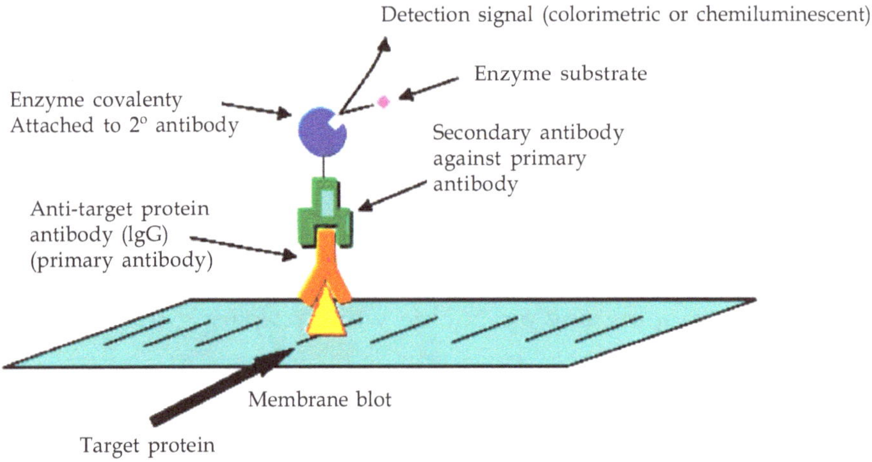

Fig. 25: Concept of Western Blot

Section-7

Plant Biotechnology

Chapter 42

Plant Tissue Culture Techniques

Media Preparation

- Before mixing medium, determine the quantity of each medium and the amount of plant growth regulator needed for each experiment. Make slightly more medium than is required. A media check off list is helpful in making medium to keep track of ingredients.
- Remove media stock solutions and plant growth regulator stock solutions to be used from the refrigerator. Make up fresh stock solutions if adequate quantities are not available or if solutions have precipitated.
- Plant growth regulators stock solutions are prepared biweekly and stored at 4° C in the dark. Cytokinins (thidiazuron and BA) are dissolved in approximately KOH prior the addition of deionized water. Gibberellins and auxins (2,4-D, IBA and NAA) are dissolved in 100% ethanol prior the addition of water.
- Fill the volumetric flasks upto ½ to ¾ with deionized water. Add proper a of stock solution and plant growth regulators and mark

each item on the check off list. Add 30 g/l sucrose, stir until dissolved. Make up the volume of the flask with deionized water and later por into a larger beaker.

- Calibrate the pH to 5.8 with 1 N KOH or HC1 with a pH meter prior to adjusting the pH of the medium.
- Add Sigma agar, 1g/l, to the medium. Place the medium on high heat and stir until the agar dissolves. The medium will appear translucent when the agar is cpmpletely dissolved.
- Pour proper aliquots into the culture vessels; test tubes should contain 15 ml medium; glass baby food jars 25 ml. Place capped culture vessels in the autoclave for 20 minutes at 121° C and 1 kg cm^{-2}.
- Store the autoclaved medium on the clean medium racks in the laboratory. Avoid storing medium for more than 2 weeks, it is better to make fresh medium for successful, repeatable experiments.

Special Preparation for Medium Containing Antibiotics

Medium containing filtered sterilized ingredients (antibiotics) is prepared differently since autoclaving may destroy antibiotic.

- Dissolve Antibiotics in water or phosphate buffer (pH 7.0). Filter the stock solutions through a 22 µm sterilized filter disk attached to a syringe into sterile glass baby food jars under sterile conditions in a laminar flow hood. Store the filtered sterilized stock solutions in sterile baby food jars at 0° C in the dark.
- Medium preparation is the same as above except after the addition of agar, the medium is directly placed into Borosil flasks and autoclaved for 20 minutes. Do not fill the flasks more than 2/3 to 3/4 full with medium.
- Remove the medium from the autoclave. When the temperature reaches approximately 60°C, add the filtered sterilized ingredients to the medium under sterile conditions in a laminar flow hood. After mixing, pour the medium into sterile culture vessels.
- Store the medium at 4°C in the dark for no longer than 2 weeks.

Chapter 43

In vitro Germination of Orchid Seeds

Principle

Orchid seeds are produced in large numbers but are exceedingly small, hence are also called "dust seeds" may be as small as 0.01 by 0.05 mm in dimension and weigh as little as 10-30 ug. Seedling development is generally a slow process in orchids, with the seedlings passing through an initial stage, termed the protocorm stage, prior to the production of distinct vegetative organs (leaves, roots, etc). The protocorms of terrestrial orchids are subterranean and entirely heterotrophic, with all of the nutritional needs of the orchid being supplied by the mycorrhizae (the symbiotic fungus). The length of the protocorm stage is variable, however, in many cases, months, or even years, elapse before leaves and roots are produced.

The combination of a small seed size and fungal symbiont requirement largely precludes the propagation via conventional seed propagation protocols. Accordingly, most of the early research on the development of orchid seed germination protocols focused on *in vitro* germination with the seeds and fungal symbiont being sown on the same petri plate. Successful protocols were developed for several species; however, the process was complicated by the diversity of fungal symbionts required (across orchid taxa) and the difficulty in isolating and culturing these fungi.

In the 1920's the process of orchid seed propagation was revolutionized by Lewis Knudson with the demonstration that nutrient media that could be used as a substitute for fungal infection. The isolation and culture of multiple fungal strains was no longer required, and improvements in germination frequency, germination speed, and seedling vigor were generally observed. To date, successful asymbiotic germination techniques, termed 'flasking', have been developed for many epiphytic orchids, with several species being saved from extinction via the application of these techniques.

Germination Medium

The seeds will be germinated on a MS medium with half-strength mineral salts and 20 g/L sucrose. Although MS medium was not specifically developed for orchid seed germination, it has generally been found sufficient for demonstration purposes.

Protocol

- ❒ Take orchid seeds in plastic Pasteur pipette of orchid seeds. Plug the pipettes with cotton wool to contain the seeds.
- ❒ Working in a laminar flow hood, immerse the lower two-thirds of the pipette in a 30% bleach solution. Allow 15-30 seconds for the excess bleach to drain from the pipette.
- ❒ Draw 3-4 ml of 10% bleach solution (with 0.05% Tween-20 detergent). Invert the pipette to ensure that all internal surfaces are rinsed and shake the pipette vigorously for 15-20 seconds. Repeat the shaking every 2 minutes until 10 minutes of exposure time has elapsed.
- ❒ Expel the bleach solution (being careful not to apply so much pressure as to dislodge the cotton plug) and draw approximately 3-4 mls of sterile water into the pipette. Shake or mix the pipette briefly to rinse the seeds and allow at least 60-90 seconds exposure time before expelling the water slowly from the pipette. Repeat two more times.
- ❒ After the final rinse, draw 1-1, 5 mls of water into the pipette to be used as a carrier for the seed. Remove the cotton plug using

sterile forceps and dispense the seeds onto the germination media in the petri plates and jars. Disperse the seeds with swirling and remove any excess water from the media's surface using a pipette with a sterile tip.

- After seed sowing, seal the plate with parafilm (2 xs) and place the plates and jars both in the growth chamber (16 hr photoperiod, 23 ± 2 °C).

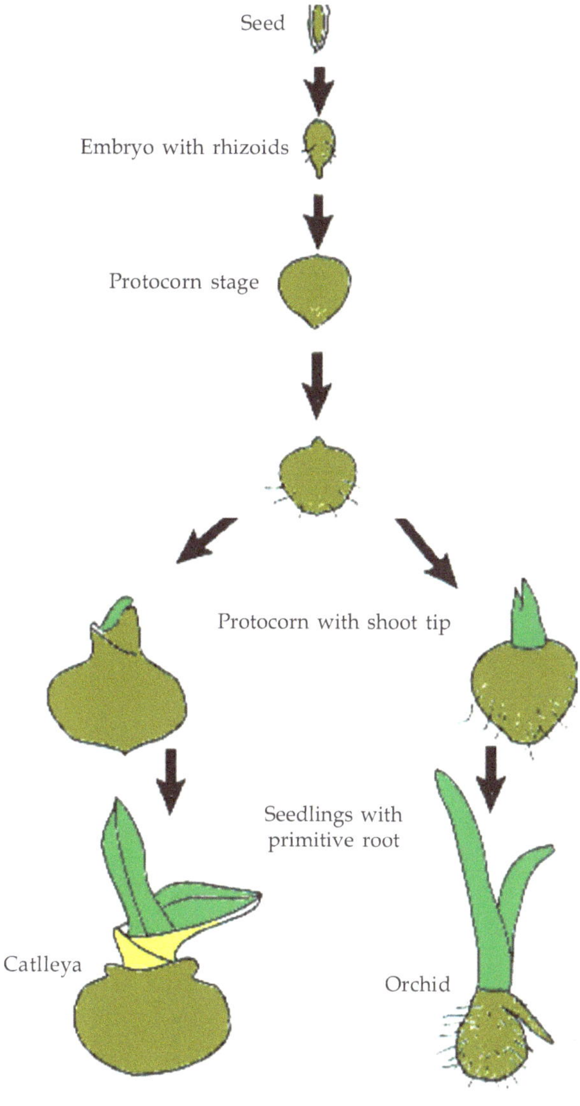

Fig. 26: *In Vitro* Germination of Orchid Seeds

- Starting after two to three weeks examine the plates on a regular basis for signs of seed germination using a dissecting microscope. Early stages of protocorm development should be apparently after 3-4 wks, with the protocorms appearing as small green balls.

  Note: The protocorms of many orchids produce tubular epidermal outgrowths, termed rhizoids that are thought to serve as a site of fugal entry during in vivo germination.

- Leaf and root production may be apparent within a matter of months. Root formation is generally thought to preceed leaf formation; however, the initial (primary) root is often small and difficult to identify. In addition the primary root is short-lived, and is ultimately replaced by adventitious roots which may be visible by the two to three leaf stages.

- Draw sketches of the different stages of seedling development observed.

  Note: In general, orchid seedlings are sub-cultured every 3-6 months until they are large enough for transplanting. However, some seedlings may grow to transplanting size without subculture.

Chapter 44

Micropropagation - Initiation of Miniature Rose Shoot Culture

Principle

Micropropagation is especially useful for include propagation of high value plants that grow slowly (bulbs, ferns, orchids, palms, etc), new cultivars having limited stock and high market demand (new cultivars produced via conventional breeding or genetic engineering), cultivars that are difficult to root by conventional techniques, and endangered plant species. In addition, virus elimination is possible via culture initiation from meristem explants. Also, micropropagation provides a source of sterile tissues that can be used for other tissue culture protocols, including genetic transformation.

Micropropagation is possible using multiple types of cultures; however, shoot cultures are typically employed, in which case five stages to the micropropagation process are typically recognized as outlined below. Nodal explants of rose are used for culture initiation. Three media, differing in hormone content, is employed and the explants should be evaluated for differences in axillary shoot growth and proliferation.

Stages of Micropropagation (Shoot Culture)

Stage 0 - Explant selection and pretreatment

Stage 1 - Establishment of aseptic cultures (sterilization) and stabilization

Stage 2 - Multiplication.

Note: In the case of shoot cultures multiplication is possible via the use of nodal explants or the induction of spontaneous axillary bud break. Axillary bud break is promoted via the use of relatively high cytokinin concentrations; however, the shoots produced are often somewhat stunted and may require transfer to a medium with somewhat lower cytokinin concentrations prior to rooting or transplanting. Shoot elongation during the multiplication is frequently facilitated by the inclusion of low concentrations of auxin (IBA, IAA, NAA, etc.). The multiplication stage is repeated until the required number of plantlets has been produced.

Stage 3 - *In vitro* rooting and conditioning for transplanting.

Note: Rooting is generally achieved using hormone free (basal) media, or media supplemented with auxins (i.e., less nutrients and sucrose, lower temperatures, stronger lights). In many cases this step is skipped with roots forming upon transfer to the soil.

Stage 4 - Transplanting to the soil (ex-vitro) and acclimization to ambient humidity.

Plant Material

- Stockplants of miniature rose.

• *Media*

1) MS medium with 20 g/L sucrose
2) MS medium with 20 g/L sucrose + 3 mg/L BA
3) MS medium with 20 g/L sucrose + 3 mg/L BA + 0.3 mg/L IBA

Note: All three media were gelled with Gelrite (3 g/L) and prepared in 25 x 150 mm glass culture tubes. For each media type, 6 replications, each consisting of a single explant, shoud be prepared.

Protocol

- Collect 20-24 single node explants from one of the stockplants, leaving 1 to 2 cm below each target node to facilitate the sticking of the explants in the media. The leaves should also be removed from the explants.

- Wash the explants well with soapy water using your fingers or a bottle brush. Remove the detergent by rinsing and place the explants in a glass baby food jar.
- Add enough 70% ethanol to cover the explants and cap the jar. Invert and mix the jar. After 30-45 seconds pour off the ethanol and replace with 20% bleach (with 0.05% Tween-20). Seal with parafilm and mix by inversion to ensure the entire inner surface of the jar is exposed to the bleach.

  Note: When changing solutions be sure to pour off the previous solution slowly. The explants may remain at the bottom of the jar. Otherwise they should be caught at the lip.

- Mix the explants by swirling every 3-5 minutes. After 15-20 minutes total exposure time, pour off the bleach and replace with sterile water
- After 3-4 distilled water rinses, the explants should be blotted dry on sterile paper towels and transferred to the experimental media using sterile forceps. The explants should be randomly assigned to the experimental media. For each medium, six replications, each with a single explant should be prepared.
- The culture tubes of explants should be placed in the walk-in growth chamber (SB Rm321, 16 hr photoperiod, 23 ± 2 °C).
- After three weeks the explants should be examined for sterility, axillary shoot growth, and possible also axillary proliferation.

Chapter 45

Potato Transformation Protocol

This protocol is designed for potato stem internodes on ZIG medium, but might work for leaf pieces as well.

Plant Material

Cut stem internode explants from 5-6-week-old *in vitro* plants (or leaf explants from 3-4-week-old *in vitro* plants) grown on 15 ml PROP in 25 x 150 mm test tubes.

Day 1

Streak out desired *Agrobacterium* strain on YEP + antibiotic to obtain a single colony. Grow at 28 C for 2 days. Alternatively, skip this step and use frozen stock for next step.

Day 3

Use a single *Agrobacterium* colony to inoculate 20 ml YEP + antibiotic in 250-ml flask. Grow overnight at 28 C with vigorous shaking (~200 rpm). Some *Agrobacterium* strains require an additional day to reach sufficient density.

Day 4

- ❒ Cut explants (either 0.5-1.0-cm stem internodes or trimmed leaf pieces) on moist filter paper; we use Fisher P8 creped filter paper. You want to make clean cuts; avoid crushing the tissue. Make sure there are no nodal regions attached to the explants. Place cut explants in a small amount of liquid MS medium. (See note below for suggested numbers.)
- ❒ Remove control pieces from liquid MS medium, blot pieces on filter paper to remove excess moisture, and place on ZIG medium (~ 16 ml per 100 x 15 mm plate). 15 explants per plate. Seal plates with parafilm and place in an incubator (60-80 $uEm^{-2}s^{-1}$, 22 C, 16-h photoperiod). Cover plates with four layers of cheesecloth to maintain extremely low light conditions.
- ❒ Estimate concentration of *Agrobacteria* by measuring absorbance at 600 nm. Dilute culture with YEP-NaCl (YEP without NaCl) until absorbance is between 0.6 and 0.8 (4.6-6.1 x 10^8 cells/ml).
- ❒ Place appropriate number of explants in diluted *Agrobacterium* solution for 30 minutes. Blot pieces on filter paper to remove excess bacteria and transfer to ZIG medium. 15 explants per plate. Seal plates with parafilm and place in an incubator (60-80 $uEm^{-2}s^{-1}$, 22 C, 16-h photoperiod). Cover plates with four layers of cheesecloth to maintain extremely low light conditions.

Day 8

Transfer explants to fresh ZIG medium (this time, ~ 33 ml per 100 X 25 mm plate, thick plates) containing appropriate antibiotics. (See note below for suggests numbers.) If *Agrobacterium* growth is excessive, you can rinse explants three times in sterile water before moving them to ZIG medium containing antibiotics. Seal plates with parafilm and cut 2 slits (1 cm long) on opposite edges of each sealed plate. Place plates in an incubator (60-80 $uEm^{-2}s^{-1}$, 22 C, 16-h photoperiod) and cover with two layers of cheesecloth to maintain low light conditions.

Option: On Day 8, include only the antibiotic that inhibits *Agrobacterium* growth. On Day 10, transfer explants to ZIG + antibiotics that inhibit *Agrobacteria* **and** select transformed cells.

Day 22

Transfer explants to fresh Zig medium (~ 33 ml per 100 X 25 mm plate) containing appropriate antibiotics every 10-16 days. Same conditions as above. When shoots appear, remove cheesecloth,

Day 36

When the shoots are ~ 2 cm long, remove them from the explants. Make a clean cut at the base of each shoot to ensure there is no callus tissue attached to it. Place each shoot in 10 ml PROP in a 20 x 150 mm test tube. You can include the antibiotic that inhibits *Agrobacteria* growth, but we have not found it necessary. We generally include the antibiotic that selects transformants; the few non-transformed shoots that form on the ZIG plates generally do not develop roots in PROP containing antibiotic.

Media

Potato Propagatioin Medium (PROP)

- ❐ Volume: 1 liter
- ❐ MS Basal Salts: 4.33 g
- ❐ MS Stock C: 5 ml
- ❐ MS Stock D: 5 ml
- ❐ Sucrose: 30 g
- ❐ Myo-inositol: 100 mg
- ❐ 1000x Prop Vitamins: 1 ml
- ❐ pH 5.8
- ❐ Agar (Difco Bacto): 8 g

MS Stock C

- ❐ Volume: 1 liter
- ❐ $CaCl_2 \cdot 2H_2O$: 88 g

MS Stock D

- ❐ Volume: 1 liter
- ❐ KH_2PO_4: 34 g

1000x Prop Vitamins

- ❐ Volume: 100 ml
- ❐ glycine: 40 mg
- ❐ nicotinic acid: 10 mg
- ❐ pyridoxine HCl: 10 mg
- ❐ thiamine HCl: 10 mg

YEP Medium (Solid)

- Volume: 100 ml
- Bacto Peptone: 1 g
- Bacto Yeast Extract: 1 g
- NaCl: 0.5 g
- Agar (Difco Bacto): 1.5 g

MS Medium (Liquid)

- Volume: 1 liter
- MS Basal Salts: 4.33 g
- sucrose: 30 g
- 1000x MS Vitamins: 1 ml
- pH 5.7

1000x MS Vitamins

- Volume: 100 ml
- glycine: 200 mg
- myo-inositol: 10 g
- nicotinic acid: 50 mg
- pyridoxine HCl: 50 mg
- thiamine HCl: 10 mg

ZIG Medium

Volume: 1 liter

- MS Basal Salts: 4.33 g
- sucrose: 20 g
- 1000x N&N Vitamins: 1 ml
- zeatin riboside 4.78 mg
- IAA 0.01 mg
- GA3 0.2 mg
- pH 5.7
- Agar (Noble) 7 g

1000x N&N Vitamins

- Volume: 100 ml
- biotin (dissolve in KOH): 5 mg

- folic acid (dissolve in KOH): 50 mg
- glycine: 200 mg
- myo-inositol: 10 g
- nicotinic acid: 500 mg
- pyridoxine HCl: 50 mg
- thiamine HCl: 50 mg

Note: When using stem internodes, a critical factor is internode diameter. The best results have been seen with diameters 2 mm or greater. With some genotypes, leaf explants (versus stem internodes) result in a higher rate of transformation. While using leaves, retain petiole. Remove distal portion of leaf to expose vascular tissue. Minimum co-cultivation time results in the lowest rate of transformation, but also the lowest amount of damage to explants. You might be able to extend the co-cultivation time by a few days.

Chapter 46

Transformation : *Agrobacterium* (Freeze Thaw Method)

After preparing the desired DNA molecule using *E.coli*, the plasmid can be transferred into Agrobacterium by this method. Efficiency is low (~10^3 transformants per mg of DNA) but the method is reliable and very rapid. This method also eliminates much of the plasmid rearrangement that occurs during triparental mating.

Procedure

- Grow an Agribacterium strain containing an appropriate helper Ti plasmid in 5 ml of LB medium 16 hours at 28°C.
- Add 2 ml of the culture to 50 ml LB medium in a 250 ml flask and shake (~250 RPM) at 28°C until an OD_{600} of 0.5 to 1.0 is obtained.
- Chill the culture on ice. Centrifuge the cell suspension at 1000 g for 30 minutes at 4°C.
- Discard the supernatant. Resuspend the cells in 1 ml of 20 mM $CaCl_2$ (ice cold). Dispense 100 mL aliquots into pre-chilled 1.5 ml tubes.

- Add about 1 μg of the desired plasmid DNA (pure or crude) to the cells in each tube.
- Freeze the cells in liquid nitrogen (dry ice - ethanol bath freezing yields lower efficiency).

  Note: The cells (less the DNA) can be frozen in liquid nitrogen and stored at -80°C for use with future transformations. Add ~ 1 μg of DNA to the frozen cells and continue with step 7.

- Thaw the cells by incubating the tubes in a 37 °C water bath for 5 minutes.

  Add I ml of LB medium to each tube and incubate at 28 C with gentle shaking for 2 to 4 hours. This allows resistance gene expression.

- Centrifuge the tubes to pellet the cells (2000 g for 5 minutes). Discard the supernatant. Resuspend the cells in 100 μL of LB medium.
- Spread the cells on LB plates containing the appropriate antibiotic(s) for the cell line and your plasmid. Incubate the plates at 28°C for 2 to 3 days until colonies appear.
- Pick isolated colonies and process as usual.

Appendices

APPENDIX-I : Solutions & Buffers

Solutions

10X ABI TBE

- ❒ 216 g Tris base
- ❒ 110 g boric acid
- ❒ 16.6 g EDTA
- ❒ Add water to 2 liters.

40% Acrylamide/Bisacrylamide (40% A&B)

- ❒ 380 g Acrylamide (Kodak 5521)
- ❒ 20 g N,N-Methylene-bisacrylamide (Kodak 8383)

Dissolve in approx. 800 ml of double distilled water and then deionize by stirring with 50 g of Amberlite MB-1 (Sigma MB-1A) for 1 hour at room temperature. Suction filter to remove the Amberlite and adjust to a final volume of 1 liter with double distilled water. (store at 4deg.C).

10x Agarose gel loading dye

15% Ficoll, 0.2% bromophenol blue, 0.2% xylene cyanol FF in double distilled water.

1.5 g	Ficoll
0.02 g	Bromophenol blue
0.02 g	xylene cyanole FF
ddH_2O to 10 ml	(store at -20°C).

Alkaline lysis solution (NaOH/SDS)

0.2 N NaOH, 1% SDS in ddwater.

- ❒ 20 ml of 1 N NaOH (or 0.8 gms)
- ❒ 10 ml of 10% SDS (or 1.0 gms)
- ❒ ddH_2O to 100 ml (make fresh)

15% Ammonium persulfate (APS)

- ❒ g APS (Kodak 11151)
- ❒ ddH_2O to 10 ml (store at 4°C).

Ampicillin (Amp)

Stock of 5 mg/ml in sterile ddwater ($sddH_2O$).

- ❒ 0.5 g Amp
- ❒ $sddH_2O$ to 100 ml (Add to media for final conc. 100 μg/ml)

100 mM rATP (adenosine triphosphate)

- ❒ 619 mg dipotassium ATP (ICN 100004)
- ❒ ddH_2O to 10 ml (aliquot and store at -20°C).

1 mg/ml BSA (bovine serum albumin)

- ❒ 5 mg BSA
- ❒ $sddH_2O$ to 5 ml (aliquot and store at -20°C)

Dilution buffer

50 mM HEPES, pH 7.6, 10 mM $MgCl_2$, 1 mM DTT, and 1 mg/ml BSA in sterile double distilled water.

500 μl	1 M HEPES, pH 7.6
100 μl	1 M $MgCl_2$

10 µl	1 M DTT
10 mg	BSA
ddH_2O to 10 ml	

Reaction buffer

500 mM Tris-HCl, pH 8.5 and 150 mM $MgCl_2$ in sterile double distilled water.

5 ml	1 M Tris-HCl, pH 8.5
1.5 ml	1 M $MgCl_2$
ddH_2O to 10 ml	

Nucleotide extension mix

15 uM dCTP, 7deaza-dGTP, and dTTP in 5 mM Tris-HCl, pH 7.6 and 0.1 mM EDTA

3 µl	5 mM dCTP
3 µl	5 mM 7deaza-dGTP
3 µl	5 mM dTTP
100 µl	50:1 TE buffer
891 µl	$sddH_2O$
Total	**1 ml**

"Short" termination "A" mix

8 µM dATP, 164 µM dCTP, 164 µM 7deaza-dGTP, 164 uM dTTP, 660 µM ddATP in 5 mM Tris-HCl, pH 7.6 and 0.1 mM EDTA.

8 µl	0.5 mM dATP
16.4 µl	5 mM dCTP
16.4 µl	5 mM 7deaza-dGTP
16.4 µl	5 mM dTTP
66 µl	5 mM ddATP
50 µl	50:1 TE buffer
326.8 µl	$sddH_2O$
Total:	**500 ul**

aliquot (18 µl for 6 reactions) and store at -70°C

"Short" termination "C" mix

8 µMCTP, 164 µM dATP, 164 µM 7deaza-dGTP, 164 µM dTTP, 400 µM ddCTP in 5 mM Tris-HCl, pH 7.6 and 0.1 mM EDTA.

16.4 μl	5 mM dATP
8 μl	0.5 mM dCTP
16.4 μl	5 mM 7deaza-dGTP
16.4 μl	5 mM dTTP
40 μl	5 mM ddCTP
50 μl	50:1 TE buffer
352.8 μl l	sddH_2O
Total	**500 μl**

aliquot (18 μl for 6 reactions) and store at -70°C

"Short" termination "G" mix

8 μM dGTP, 164 μM dATP, 164 μM dCTP, 164 μM dTTP, 540 μM ddGTP in 5 mM Tris-HCl, pH 7.6 and 0.1 mM EDTA.

16.4 μl	5 mM dATP
16.4 μl	5 mM dCTP
8 μl	0.5 mM 7deaza-dGTP
16.4 μl	5 mM dTTP
54 μl	5 mM ddGTP
50 μl	50:1 TE buffer
338.8 μl	sddH_2O
Total	**500 μl**

aliquot (18 μl for 6 reactions) and store at -70°C

"Short" termination "T" mix

8 μM dTTP, 164μM dATP, 164μM dCTP, 164μM dTTP, 600μM ddTTP in 5 mM Tris-HCl, pH 7.6 and 0.1 mM EDTA.

16.4 μl	5 mM dATP
16.4 μl	5 mM dCTP
16.4 μl	5 mM 7deaza-dGTP
8 μl	0.5 mM dTTP
60 μl	5 mM ddTTP
50 μl	50:1 TE buffer
332.8 μl	sddH_2O
Total	**500 μl**

aliquot (18 μl for 6 reactions) and store at -70°C

"Long" termination "A" mix

8 μM dATP, 164 μM dCTP, 164μM 7deaza-dGTP, 164μM dTTP, 110 μM ddATP in 5 mM Tris-HCl, pH 7.6 and 0.1 mM EDTA.

8 μl	0.5 mM dATP
16. μl	5 mM dCTP
16.4 μl	5 mM 7deaza-dGTP
16.4 μl	5 mM dTTP
11 μl	5 mM ddATP
50 μl	50:1 TE buffer
381.8 μl	sddH_2O
Total	**500 μl**

aliquot (18 μl for 6 reactions) and store at -70°C

"Long" termination "C" mix

8 μM dCTP, 164 μM dATP, 164 μM 7deaza-dGTP, 164 μM dTTP, 65 μM ddCTP in 5 mM Tris-HCl, pH 7.6 and 0.1 mM EDTA.

16.4μl	5 mM dATP
8μl	0.5 mM dCTP
16.4 μl	5 mM 7deaza-dGTP
16.4 μl	5 mM dTTP
6.5 μl	5 mM ddCTP
50 μl	50:1 TE buffer
386.3μl	sddH_2O
Total	**500 μl**

aliquot (18 μl for 6 reactions) and store at -70°C

"Long" termination "G" mix

8 μM dGTP, 164 μM dATP, 164 μM dCTP, 164 μM dTTP, 70 μM ddGTP in 5 mM Tris-HCl, pH 7.6 and 0.1 mM EDTA.

16.4 μl	5 mM dATP
16.4 μl	5 mM dCTP
8 μl	0.5 mM 7deaza-dGTP
16.4μl	5 mM dTTP
7 μl	5 mM ddGTP
50 ul	50:1 TE buffer
385.8 μl	sddH_2O
Total	**500 μl**

aliquot (18 vl for 6 reactions) and store at -70°C

"Long" termination "T" mix

8 μM dTTP, 164μM dATP, 164 μM dCTP, 164 μM dTTP, 150 μM ddTTP in 5 mM Tris-HCl, pH 7.6 and 0.1 mM EDTA.

16.4 μl	5 mM dATP
16.4 μl	5 mM dCTP
16.4 μl	5 mM 7deaza-dGTP
8 μl	0.5 mM dTTP
15μl	5 mM ddTTP
50μl	50:1 TE buffer
377.8μl	sddH_2O
Total	**500 μl**

aliquot (18 μl for 6 reactions) and store at -70°C

100 mM calcium chloride ($CaCl_2$)

- 1.48 g $CaCl_2$-$2H_2O$
- ddH2O to 100 ml
- autoclave to sterilize (store at 4°C).

50 mM calcium chloride

- 0.74 g $CaCl_2$-$2H_2O$
- ddH_2O to 100 ml
- autoclave to sterilize (store at 4deg.C).

Chloramphenicol 10 mg/ml stock

- 2 g Chloramphenicol
- 200 ml 100% (or 95%) ethanol
- Final volume 200 ml
- Sterilize by filtering
- Store in the cold room in a foil wrapped screw cap borosil bottle
- Use as an additive to microbial growth media at a final concentration of 15 μg/ml media.
- For 200 ml media, add 300 μl of 10 mg/ml chloramphenicol stock.
- For 1 liter media, add 1.5 ml of 10 mg/ml chloramphenicol stock.

Deionized formamide

Stir formamide (Schwarz/Mann Biotech 800686) with Amberlite MB resin, 10 g. per 100 ml, for one hour to deionize; filter through Whatman 3MM paper, store in a dark bottle at room temperature or 4°C.

10X denaturing buffer

200 mM Tris-HCl, pH 9.5, 1 mM EDTA, 10 mM spermidine in double distilled water.

2 ml	1 M Tris-HCl, pH 9.5
20 ul	0.5 M EDTA, pH 8.0
1 ml	100 mM spermidine
DdH_2O to 10 ml (aliquot and store at -20°C)	

Diatomaceous earth (100 mg/ml)

Suspend 10 g of diatomaceous earth (Sigma D-5384) in 100 ml of distilled water in a 100 ml graduated cylinder, and let it settle down for 3 hours. Decant the supernatant, and resuspend the pellet in 100 ml of 6 M guanidine HCl, pH 6.4, containing 50 mM Tris-HCl, 20 mM EDTA.

Diatomaceous earth-wash buffer

10 mM Tris-HCl, pH 8.0, 1 mM EDTA, pH 8.0, 50% ethanol in double distilled water.

10 ml	1 M Tris-HCl, pH 8.0
2 ml	0.5 M EDTA, pH 8.0
500 ml	100% ethanol
ddH_2O to 1 L	

1 M DTT (Dithiothreitol, Cleland's reagent)

- 1.54 g DTT (Calbiochem 233155)
- ddH_2O to 10 ml (aliquot and store at -20°C).

DNase-free RNase A

20 mg/ml RNase A in 1 mM NaOAc, pH 4.5.

200 mg	RNase A (Sigma R-5500)
3.3 ul	3 M NaOAc, pH 4.5
ddH_2O to 10 ml	

boil for 10 minutes (aliquot and store at -20°C).

0.5 M EDTA, pH 8.0 (disodium ethylenediamine tetraacetate)

186.1 g	Na_2EDTA

Dissolve in approx. 400 ml ddH_2O, adjust pH to 8.0 with 10 N NaOH, and adjust to 1 liter final volume with distilled water.

100 mM EDTA

20 ml	0.5 M EDTA
80 ml	ddH_2O
Total	**100 ml**

25mM EDTA, 50 mg/ml Blue Dextran - ABI377 Loading Dye/ Formamide mixture

Add 0.93g of EDTA to 90 ml water. Then, adjust the pH to 8.0. Bring the final volume to 100 ml. Next, add 50 mg Blue Dextran to a 1 ml EDTA solution. Add 1 ul of a 1:5 solution of this loading dye:deionized formamide to each sample well for loading onto the ABI377.

95% ethanol/0.12 M NaOAc (ethanol/acetate)

95 ml	100% ethanol
4 ml	3 M NaOAc pH 4.5
1 ml	ddH_2O
Total	**100 ml**

5 mg/ml ethidium bromide (EtBr)

- 500 mg EtBr (Sigma E-8751)
- ddH_2O to 100 ml

10X Freezer Media (FM) for storing either shotgun plasmid-based sub-clone or cDNA clones in microtiter plates)

Final Concentration	1L	500 ml	250 ml
360 mM K_2HPO_4	62.7 gm	31.35 gm	15.68 gm
132 mM KH_2PO_4	17.96 gm	8.98 gm	4.49 gm
17 mM Sodium Citrate	5.0 gm	2.5 gm	1.25 gm
4 mM $MgSO_4.7H_2O$	0.98 gm	0.49 gm	0.24 gm
(or 1M MgSO4	4 ml	2 ml	1 ml)
68 mM $(NH_4)_2SO_4$	8.98 gm	4.49 gm	2.25 gm
44% Glycerol	440 ml	220 ml	110 ml

Bring to volume with dH_2O
Sterilize by filtration throught 0.2um filter

Then the final growth and storage media is prepared in a ratio of 9 volumes LB media and 1 volume of this 10X Freezer Media (FM).

FE (formamide/EDTA)

5:1 (v/v) formamide:50 mM EDTA

10 µl	ddH_2O
10µl	100 mM EDTA
100µl	deionized formamide

make fresh

10X Fill-in/Kinase buffer

500 mM Tris-HCl, pH 7.6, 100 mM $MgCl_2$, 10 mM DTT, and 50 µg/ml BSA in double distilled water.

5 ml	1 M Tris-HCl, pH 7.6
1 ml	1 M $MgCl_2$
100 ul	1 M DTT
500 ul	1 mg/ml BSA
3.4 ml	ddH_2O
Total	**10 ml**

Fill-in Deoxynucleotide Preparation

To make 4 ml of the fill-in nucleotides at a concentration of 0.25 mM of each nucleotide, combine the following:

- 500 µl PCR dNTPs (2 mM)
- 3500µl ddH_2O
- Aliquot this into 0.5 ml eppendorf tubes with 10 ul in each tube.

Make 4 ml of these nucleotides at a final concentration of 0.25 mM from the stock 100 mM solutions, add the following:

10 µl	100 mM dATP
10 µl	100 mM dCTP
10 µl	100 mM dGTP
10 µl	100 mM dTTP
3.6 ml	ddH_2O

Aliquot into 0.5 eppendorf tubes with 10 ul in each tube.

20% glucose

- 20 g d-glucose
- ddH_2O to 100 ml
- filter sterilize

Sterile glycerin (sterile glycerol)

Either autoclave the bottle of ACS grade Glycerin with loosened cap after putting autocalve tape on the bottle (to indicate that it has been autoclaved) or we make a 50% solution of the same Glycerin with ultrapure water (which then is easier to work with) and autoclave that in a convenient size bottle.

6 M guanidine HCl, pH 6.4, containing 50 mM Tris-HCl, 20 mM EDTA

573.18 g	Guanidine-HCl (Sigma G-4505)
50 ml	1 M Tris-HCl, pH 7.6
40 ml	0.5 M EDTA, pH 8.0
ddH_2O to	1 liter

GET/lysozyme solution

50 mM glucose, 25 mM Tris-HCl, pH 8.0, and 10 mM EDTA, pH 8.0 in double distilled water,

0.9 g	d-glucose
2.5 ml	1 M Tris-HCl, pH 8.0
2 ml	0.5 M EDTA, pH 8.0

ddH_2O to 100 ml (filter sterilize and store at 4°C).

Add 2 mg/ml lysozyme just before use.

1 M HEPES, pH 7.5

- 23.83 g HEPES
- ddH2O to 100 ml
- adjust pH to 7.5 with potassium hydroxide (KOH) (store at 4°C).

IPTG (isopropyl b-D-thiogalactopyranoside)

25 mg/ml IPTG in double distilled water

- 250 mg IPTG
- ddH_2O to 10 ml (aliquot and store at -20°C)

1 M Isocitrate (sodium salt-dihydrate)

- 29.41 g Na_3isocitrate-$2H_2O$
- ddH2O to 100 ml

10x Kinase buffer

500 mM Tris-HCl, pH 7.6, 100 mM MgCl2, and 100 mM DTT in sterile double distilled water.

5 ml	1 M Tris-HCl, pH 7.6
1 ml	1 M $MgCl_2$
1 ml	1 M DTT

ddH_2O to 10 ml (store in 25 ml aliquots at -20°C).

Kanamycin sulfate (Kan)

Stock of 5 mg/ml in sterile double distilled water (ddH_2O).

- 0.5 g Kanamycin
- $sddH_2O$ to 100 ml (Add to media for final conc. 20 ug/ml)

1M KCl (potassium chloride)

- 7.5 g KCl
- ddH_2O to 100 ml

Lambda plates

10 g	Bacto-tryptone
15 g	Bacto-agar
2.5 g	NaCl

ddH_2O to 1 L

autoclave to sterilize and pour into sterile petri dishes (approx. 20 ml/plate).

Lambda top agar

10 g	Bacto-tryptone
10 g	Bacto-agar
5 g	NaCl
ddH_2O to	1 L

autoclave to sterilize

LB Medium

10 g	Bacto-Tryptone
5 g	Bacto-yeast extract
10 g	NaCl
ddH_2O to	1 L

adjust the pH to 7.0 and then autoclave to sterilize

LB plates

10 g	Bacto-Tryptone
5 g	Bacto-yeast extract
10 g	NaCl
15 g	Bacto-agar
ddH_2O to	1 L

autoclave to sterilize, cool to 55deg.C, add antibiotic if desired, and pour into sterile petri dishes (approx. 20 ml/plate).

10x Ligation buffer

50 mM Tris-HCl, pH 7.6, 10 mM MgCl2, 10 mM DTT, 10 mM rATP, and 250µg/ml bovine serum albumin (BSA) in sterile double distilled water.

5 ml	1 M Tris-HCl, pH 7.6
1 ml	1 M $MgCl_2$
1 ml	1 M DTT
1 ml	100 mM rATP
2.5 mg	BSA
$sddH_2O$ to	10 ml (store in 25 ml aliquots at -20°.C)

Loading dye

0.3% xylene cyanole FF, 0.3% bromophenol blue, 10 mM EDTA in deionized formamide.

3 g	xylene cyanole FF
3 g	bromophenol blue
0.2 ml	0.5 M EDTA

deionized formamide to 10 ml

Lysozyme solution

50 mM Tris-HCl, pH 8.0, 10 mM EDTA, and 5 mg/ml lysozyme in sterile double distilled water.

5 ml	1 M Tris-HCl, pH 8.0
2 ml	0.5 M EDTA
0.5 g	lysozyme

ddH2O to 100 ml (make fresh)

M-9 Agar

Add 15 g. agar to 870 ml. double distilled water in a 2 L borosil flask and autoclave. Also autoclave a 100 ml. graduate cylinder capped with aluminum foil to use for measuring the sterile M-9 salts later. Swirl the agar gently and carefully upon removal from autoclave to disperse any undissolved agar. Allow to cool in a 55°C water bath. When 55°C, add the ingredients called for in the M-9 liquid medium recipe, omitting the water. Be sure to use sterile pipettes or graduate cylinders, as this mixture cannot be autoclaved. Immediately pour into sterile Petri dishes, using sterile technique.

M-9 Medium (liquid)

100 ml	10X M-9 salts
1 ml	1 M $MgSO_4$ (autoclaved)
10 ml	20% glucose (filter sterilized),
1 ml	1% thiamine (filter sterilized)
10 ml	100 mM $CaCl_2$ (autoclaved)
ddH_2O to	1 L

10X M-9 Salts

60 g	Na_2HPO_4 (sodium phosphate, dibasic)
30 g	KH_2PO_4 (potassium phosphate, monobasic)
5 g	NaCl
10 g	NH_4Cl (ammonium chloride)
ddH_2O to	1 liter (autoclave)

1 M MgCl2 (magnesium chloride)

- 20.33 g $MgCl_2$-6H_2O
- ddH_2O to 100 ml

1 M MgSO4 (magnesium sulfate)

12.04 g $MgSO_4$

ddH_2O to 100 ml (autoclave)

1 M MnCl2 (manganese chloride)

- 1.98 g $MnCl_2$
- ddH_2O to 10 ml (store protected from light)

1 M MOPS

- 20.93 g MOPS
- Dissolve in 80 ml ddH_2O, adjust pH to 7.5 with 1 N NaOH, and bring volume to 100 ml.

10X MOPS buffer

400 mM MOPS, pH 7.5, 500 mM NaCl, 100 mM $MgCl_2$ in double distilled water.

400 µl	1 M MOPS, pH 7.5
170 µl	3 M NaCl
100 µl	1 M $MgCl_2$
330 µl	ddH_2O
Total:	**1 ml**

2.7 M MOPS (acid form)

- 5.65 g MOPS (acid form)
- ddH_2O to 10 ml

MOPS-Acid buffer

1.35 M MOPS (acid form), 100 mM $MgCl_2$ in double distilled water.

500 µl	2.7 M MOPS (acid form)
100 µl	1 M $MgCl_2$
400 µl	ddH_2O
Total	**1 ml**

10X Mn2+/isocitrate buffer

50 mM $MnCl_2$, 150 mM isocitrate (Na salt), 25% glycerol in double distilled water

50 µl	1 M $MnCl_2$
150 µl	1 M isocitrate
250 µl	glycerol
550 µl	ddH_2O
Total	**1 ml**

10x MTBE (Modified Tris-borate-EDTA buffer)

1.3 M Tris, 0.4 M Boric acid, and 0.02 M EDTA in double distilled water.

162 g	Tris base
27.5 g	Boric acid
9.3 g	Na_2EDTA
ddH_2O to	1 L

20 mM dNTP stocks

Prepare from 100 mM stocks

80 µl	100 mM dNTP
40 µl	50:1 TE buffer
280 µl l	ddH_2O
Total	**400 µl**

5 mM dNTP stocks

Prepare from 20 mM stocks

25 µl	20 mM dNTP
10 µl	50:1 TE buffer
65 µl	ddH_2O
Total:	100 µl

2 mM dNTPs

2 mM dATP, dCTP, dGTP, and dTTP in 5 mM Tris-HCl, pH 7.6, 0.1 mM EDTA

100 µl	20 mM dATP
100 µl	20 mM dCTP
100 µl	20 mM dGTP
100 µl	20 mM dTTP
100 µl	50:1 TE buffer
500 µl	ddH_2O
Total	**1 ml**

2 mM [alpha]-S-dNTPs

2 mM [alpha]-S-dATP, [alpha]-S-dCTP, [alpha]-S-dGTP, and [alpha]-S-dTTP in 5 mM Tris-HCl, pH 7.6, 0.1 mM EDTA

100 µl	20 mM [alpha]-S-dATP
100 µl	20 mM [alpha]-S-dCTP
100 µl	20 mM [alpha]-S-dGTP
100 µl	20 mM [alpha]-S-dTTP
100 µl	50:1 TE buffer
500 µl	ddH_2O
Total	**1 ml**

3M NaCl (sodium chloride)

- ❐ 17.53 g NaCl
- ❐ ddH_2O to 100 ml

10N NaOH (sodium hydroxide)

- ❐ 40 g NaOH
- ❐ ddH_2O to 100 ml.

1N NaOH

- ❐ 10 ml 10 N NaOH
- ❐ ddH_2O to 100 ml

9.5M NH4OAc (ammonium acetate)

- ❐ 73.23 g NH4OAc
- ❐ ddH_2O to 100 ml

8.0M NH4OAc

- ❐ 61.69 g NH4OAc
- ❐ ddH_2O to 100 ml

10X PCR buffer

500 mM KCl, 100 mM Tris-HCl, pH 8.5, 15 mM $MgCl_2$ in sterile double distilled water

5 ml	1 M KCl
1 ml	1 M Tris-HCl, pH 8.5
150 µl	1 M $MgCl_2$
ddH_2O to 10 ml	

PCR Deoxynucleotide Preparation

To make 12.5 ml of the PCR nucleotides at a concentration of 2 mM each nucleotide, combine the following:

- ❒ 250 μl 100 mM dATP
- ❒ 250 μl 100 mM dCTP
- ❒ 250 μl 100 mM dGTP
- ❒ 250 μl 100 mM dTTP
- ❒ 11.5 ml ddH_2O

Aliquot this into 25 tubes with 500 ul in each tube. This will keep the nucleotides from being frozed and thawed too much.

20% PEG/2.5 M NaCl

- ❒ 7.3 g NaCl
- ❒ 10 g PEG (MW=8000)

Dissolve in 40 ml double distilled water by stirring, and then adjust the volume to 50 ml.

50% PEG/0.5 M NaCl

- ❒ 5.85 g NaCl
- ❒ 100 g PEG (MW=8000)

Dissolve in 100 ml double distilled water by stirring, and then adjust the volume to 200 ml.

PEG:TE rinse solution

1:3 solution of 20% PEG containing 2.5M NaCl and 10 mM Tris-HCl, pH 8.0 containing 1 mM EDTA in double distilled water.

250 μl	1 M Tris-HCl, pH 8.0
50 μl	0.5 M EDTA
12.5 ml	20% PEG/2.5 M NaCl.
ddH_2O to	37.5 ml

Phenol, TE-saturated

Add an equal volume of 10 mM Tris-HCl, pH 7.5-8.0, 1 mM Na2EDTA to ultrapure phenol, mix well, allow phases to separate, remove and discard upper (aqueous) phase. Repeat until the pH of the aqueous phase is between 7.5-8.0 (store at 4°C).

Phenol/chloroform/isoamyl alcohol (25:25:1)

100 ml	TE-saturated phenol
100 ml	chloroform
4 ml	isoamyl alcohol
Total	**204 ml**

2M NaOAc (sodium acetate)

- 27.22 g NaOAc-3H_2O
- ddH_2O to 100 ml

3M NaOAc, pH 4.5

- 408.24 g NaOAc-3H_2O

Dissolve in approx. 800 ml ddH_2O , adjust pH to 4.8 with glacial acetic acid and bring to a final volume of 1 L with ddH_2O.

Restriction enzyme assay buffer, 10X Low Salt

100 mM Tris-HCl, pH 7.6, 100 mM $MgCl_2$, and 10 mM DTT in sterile double distilled water.

1 ml	1 M Tris-HCl, pH 7.6
1 ml	1 M $MgCl_2$
0.1 ml	1 M DTT
ddH_2O to	10 ml

Restriction enzyme assay buffer, 10X Medium Salt

500 mM NaCl, 100 mM Tris-HCl, pH 7.6, 100 mM $MgCl_2$, and 10 mM DTT in sterile double distilled water.

1.7 ml	3 M NaCl
1 ml	1 M Tris-HCl, pH 7.6
1 ml	1 M $MgCl_2$
0.1 ml	1 M DTT
ddH_2O to	10 ml

Restriction enzyme assay buffer, 10X High Salt

1M NaCl, 500 mM Tris-HCl, pH 7.6, 100 mM $MgCl_2$, and 10 mM DTT in sterile double distilled water.

3.3 ml	3 M NaCl
5 ml	1 M Tris-HCl, pH 7.6
1 ml	1 M $MgCl_2$
0.1 ml	1 M DTT
ddH_2O to	10 ml

Restriction enzyme assay buffer, 10X SmaI

200 mM KCl, 100 mM Tris-HCl, pH 7.6, 100 mM $MgCl_2$, and 10 mM DTT in sterile double distilled water.

2 ml	1 M KCl
1 ml	1 M Tris-HCl, pH 7.6
1 ml	1 M $MgCl_2$
0.1 ml	1 M DTT
ddH_2O to	10 ml

RNase T1: 100 U/ul in 50 mM Tris-HCl, pH 7.6

100 µl	RNase T1
25 µl	1 M Tris-HCl, pH 7.6
375 µl	ddH_2O
Total	**500 µl**

10% SDS (sodium dodecyl sulfate)

- 10 g SDS
- ddH_2O to 100 ml

1X STB buffer

25% sucrose and 50 mM Tris-HCl, pH 8.0 in double distilled water.

25 g	sucrose
5 ml	1 M Tris-HCl, pH 8.0

ddH_2O to 100 ml (filter sterilize and store at 4°C)

Silanizing reagent

- 5% solution of dichloro dimethyl silane in 1,1,1-trichloroethane.

20X SSC (standard saline-citrate)

17.53 g	NaCl
8.82 g	sodium citrate

Dissolve in approx. 80 ml ddH_2O, adjust pH to 7.0 with hydrochloric acid (HCl) and bring final volume to 100 ml.

1X SSC (standard saline-citrate)

5 ml	20X SSC
95 ml	ddH_2O
Total	**100 ml**

5X Taq reaction buffer

400 mM Tris-HCl, pH 9.0, 100 mM ammonium sulfate ($(NH_4)_2SO_4$), pH 9.0, 25 mM $MgCl_2$, and 5% dimethyl sulfoxide (DMSO) in sterile double distilled water.

16 ml	1 M Tris-HCl, pH 9.0
4 ml	1 M $(NH_4)_2SO_4$, pH 9.0
1 ml	1 M $MgCl_2$
2 ml	DMSO
17 ml	ddH_2O
Total	**40 ml**

5X Taq dilution buffer

400 mM Tris-HCl, pH 9.0, 100 mM $(NH_4)_2SO_4$, pH 9.0, and 25 mM $MgCl_2$ in sterile double distilled water.

16 ml	1 M Tris-HCl, pH 9.0
4 ml	1 M $(NH_4)_2SO_4$, pH 9.0
1 ml	1 M $MgCl_2$
19 ml	ddH_2O
Total	**40 ml**

5X Taq "A" termination mix

62.5 μM dATP, 250 μM dCTP, 375 μM c7dGTP, 250 μM dTTP and 1.5 mM ddATP in 5 mM Tris-HCl, pH 7.6 and 0.1 mM EDTA.

20 μl	20 mM dATP
80 μl	20 mM dCTP
240 μl	10 mM 7deaza-dGTP
80 μl	20 mM dTTP
1920 μl	5 mM ddATP
640 μl	50:1 TE buffer
3420 μl	$sddH_2O$
Total	**6.4 ml**

5X Taq "C" termination mix

250 µM dATP, 62.5 µM dCTP, 375 µM c7dGTP, 250 µM dTTP and 0.75 mM dATP in 5 mM Tris-HCl, pH 7.6 and 0.1 mM EDTA.

80 µl	20 mM dATP
20 µl	20 mM dCTP
240 µl	10 mM 7deaza-dGTP
80 µl	20 mM dTTP
960 µl	5 mM ddCTP
640 µl	50:1 TE buffer
4380 µl	sddH_2O
Total	**6.4 ml**

5X Taq "G" termination mix

250 uM dATP, 250 uM dCTP, 94 uM c7dGTP, 250 uM dTTP and 0.125 mM ddGTP in 5 mM Tris-HCl, pH 7.6 and 0.1 mM EDTA.

160 µl	20 mM dATP
160 µl	20 mM dCTP
120 µl	10 mM 7deaza-dGTP
160 µl	20 mM dTTP
320 µl	5 mM ddGTP
1280 µl	50:1 TE buffer
10600 µl	sddH_2O
Total	**12.8 ml**

5X Taq "T" termination mix

250 uM dATP, 250 uM dCTP, 375 uM c7dGTP, 62.5 uM dTTP and 1.25 mM ddTTP in 5 mM Tris-HCl, pH 7.6 and 0.1 mM EDTA.

160 µl	20 mM dATP
160 µl	20 mM dCTP
480 µl	10 mM 7deaza-dGTP
40 µl	20 mM dTTP
3200 µl	5 mM ddTTP
1280 µl l	50:1 TE buffer
7480 µl	sddH_2O
Total	**12.8 ml**

20X TAE buffer

0.8 M Tris, 0.4 M NaOAc, and 0.04 M Na_2EDTA, and glacial acetic acid to pH 8.3 in double distilled water.

96.9 g	Tris base
32.8 g	$NaOAc\text{-}3H_2O$
14.9 g	Na_2EDTA

Dissolve in approx. 700 ml of double distilled water, adjust the pH to 8.3 with glacial acetic acid, and bring to 1 L with ddH_2O.

TEMED (N,N,N',N'-tetramethylethylenediamine)

Store protected from light at 15°C.

10xTB Salts

2.31 g	KH_2PO_4
12.54 g	K_2HPO_4 (potassium phosphate, dibasic)

ddH_2O to 100 ml autoclave)

Terrific Broth (TB)

12 g	Bacto-tryptone
24 g	yeast extract
4 ml	glycerol

ddH_2O to 900 ml

Autoclave, cool and add 100 ml of 10xTB Salts and adjust the final volume to 1 L with $sddH_2O$.

TE (10:0.1) buffer

10 mM Tris-HCl, pH 7.6, 0.1 mM EDTA

10 ml	1 M Tris-HCl, pH 7.6
0.2 ml	0.5 M EDTA

ddH_2O to 1 L

TE (10:1) buffer

10 mM Tris-HCl, pH 7.6, 1 mM EDTA

10 ml	1 M Tris-HCl, pH 7.6
2 ml	0.5 M EDTA

ddH_2O to 1 L

TE (100:10) buffer

100 mM Tris-HCl, pH 7.6, 10 mM EDTA

100 ml	1 M Tris-HCl, pH 7.6
20 ml	0.5 M EDTA
ddH_2O to 1 L	

TE (50:1) buffer

50 mM Tris-HCl, pH 7.6, 1 mM EDTA

0.5 ml	1 M Tris-HCl, pH 7.6
0.1 ml	100 mM EDTA
9.4 ml	ddH_2O
Total	**10 ml**

TE-RNase solution

50:10 TE buffer containing 40 ug/ml RNase A and 40 U/ml RNase T1

1.2 ml	1 M Tris-HCl, pH 7.6
480 µl	0.5 M EDTA
50 µl	20 mg/ml RNase A
10 µl	100U/ul RNase T1
22.3 ml	ddH_2O
Total	**24.0 ml**

Tetracycline stock (Tet)

Stock of 10 mg/ml in 50% ethanol + sddwater.

1 g	Tetracycline
50 ml	100% ethanol

$sddH_2O$ to 100 ml (store at 4°C in the absence of light)

Add to media for final conc. 20 ug/ml.

1% thiamine

- 100 mg thiamine
- $sddH_2O$ to 10 ml (filter sterilized)

1M Tris-HCl, pH 7.6, 8.0, 8.5, 9.0, 9.5

- 121.1 g Tris base
- ddH_2O to 800 ml

Adjust pH with concentrated HCl and then add ddH_2O to 1 L.

10X TBST Buffer Recipe

To make 10x TBST (wash buffer), add the following together:

- 12.11 g Tris
- 87.66 g NaCl
- 5 mL Tween-20
- 1 g Na-azide

10X TM buffer

500 mM Tris-HCl, pH 8.0, 150 mM $MgCl_2$ in sterile double distilled water.

5 ml	1 M Tris-HCl, pH 8.0
1.5 ml	1 M $MgCl_2$
ddH_2O to 10 ml	

50:2:10 TTE

50 mM Tris-HCl, pH 8.0, 2% Triton X-100, and 10 mM EDTA in double distilled water.

5 ml	1 M Tris-HCl, pH 8.0
2 ml	0.5 M EDTA
2 ml	Triton X-100 (Sigma X-100)
ddH_2O to 100 ml	

TTE

10 mM Tris-HCl, pH 8.0, 0.5% Triton X-100, and 0.1 mM EDTA in double distilled water.

500 µl	1 M Tris-HCl, pH 8.0
250 µl	Triton X-100 (Sigma X-100)
10 µl	0.5 M EDTA
ddH_2O to 50 ml	

X-gal (5-bromo-4-chloro-3-indolyl b-D-galactopyranoside)

- 200 mg x-gal (Sigma B-4252)
- dimethylformamide (DMF) to 10 ml
- Aliquot and store protected from light at -20°C)

2xTY medium

16 g	Bacto-tryptone (Difco 0123-01-1)
10 g	Bacto-yeast extract (Difco 0127-05-3)
5 g	NaCl

DdH2O to 1 L (autoclave)

Taq Cycle Sequencing Reagent Preparation

1. 5X Taq Reaction buffer

400 mM Tris-HCl, pH 9.0	16 ml	1 M Tris-HCl,	pH 9.0
100 mM $(NH_4)_2SO_4$, pH 9.0	4 ml	1 M $(NH_4)_2SO_4$,	pH 9.0
25 mM $MgCl_2$, pH 7.0	1 ml	1 M $MgCl_2$,	pH 7.0
5% DMSO	2 ml DMSO	17 ml	ddH_2O
Total	**40 ml**		

The 5X Taq reaction buffer will be added separately with the A, C, G, and T nucleotide mixes for ease in reaction pipetting. One 40 ml preparation of buffer will be sufficient for one batch (about 200 tubes) of A, C, G, and T mix aliquots.

2. Taq Dilution Buffer

400	mM Tris-HCl, pH 9.0	16 ml	1 M Tris-HCl,	pH 9.0
100	mM $(NH_4)_2SO_4$, pH 9.0	4 ml	1 M $(NH_4)_2SO_4$,	pH 9.0
25	mM $MgCl_2$, pH 7.0	1 ml	1 M $MgCl_2$,	pH 7.0
		19 ml	ddH_2O	
Total		**40 ml**		

This is routinely distributed into 30 ul aliquots in clear, unlabeled 0.5 ml microcentrifuge tubes (about 200 per batch).

3. 50:1 TE

50 mM Tris-HCl, pH 7.6	0.5 ml 1 M Tris-HCl, pH 7.6
1 mM Na_2EDTA, pH 8.0	0.1 ml 0.1 M Na_2EDTA, pH 8.0
	9.6 ml ddH_2O
Total	**10 ml**

4. Fluorescent Labeled Primers

Prepare a 100 X stock solution (40 μM); an example calculation for a dry tube of an 18mer with an O.D. of 1.00 is shown below (remembering that Joe is the dye-labeled primer for the A reaction, Fam is for C, Tamra is for G, and Rox is for T):

- OD(37μg/OD)(mol x mer/320 g)(10+12pmol/mol)(g/10+6μg)
- (1/18 mer)(μl/40 pmoles)=x ul

In this example x=160 μl, and 160 μl of ddH_20 should be added to the dried tube of fluorescent primer for a concentration of 40 uM (40 pmol/ μl). From this 100X stock of 40 uM make 1:100 dilutions. To make the amount of primer aliquoted the same as the amount of mixes per batch:

Dilute either fluorescent forward or reverse primers as follows:

64 μl of 40 uM A or C primer	128 μl of 40 uM G or T primer
6.3 ml ddH_2O	12.7 ml ddH_2O
6.4 ml	12.8 ml

For the A and C primers, distribute the 1X (0.4 μM solution) into 30 ul aliquots, and for the G and T primers, distribute them into 60 ul aliquots. The primers aliquots are stored in clear 0.5 ml microcentrifuge tubes which are labeled with blue, green, red, or yellow markers for A, C, G, or T primers, respectively.

Note: The current primers work optimally at the effective concentration of 0.4 μM, however with each new fluorescent primer preparation, the optimal concentration must be determined). The primers should be stored at 20 or -70°C.

5. 5X Taq Cycle Sequencing Mixes

Working dilutions of 20 mM are made for dATP, dCTP, and dTTP based on using one complete tube of 20 mM stock per batch of mixes. 7deaza-dGTP is purchased at a concentration of 10 mM (1080 ul are needed for one batch each of A, C, G, and T mixes, so slightly more than five tubes will be needed-each tube contains 200 μl).

	20 mM dATP	20 mM dCTP	20 mM dTTP
	95 ul of 100 mM dATP	95 ul of 100 mM dCTP	80 ul of 100 mMdTTP
	47.5 ul of 50:1 TE	47.5 ul of 50:1 TE	40 ul of 50:1 TE
	332.5 ul ddH_2O	332.5 ul ddH_2O	280 ul ddH_2O
Total	**475 μl**	**475 μl**	**400 μl**

The concentration of deoxy and dideoxy nucleotides in the mixes are shown below, followed by the recipe for one 200 tube batch of each of the four mixes.

	A	C	G	T
dATP	62.5 μM	250 μM	250 μM	250 μM
dCTP	250 μM	62.5 μM	250 μM	250 μM
7-dGTP	375 μM	375 μM	94 μM	375 μM
dTTP	250 μM	250 μM	250 μM	62.5 μM
ddATP	1.5 mM	–	–	–
ddCTP	–	0.75 mM	–	–
ddGTP	–	–	0.125 mM	–
ddTTP	–	–	–	1.25 mM

For one batch (200 tubes) of each nucleotide mix

	A	C	G	T
20 mM dATP	20 μl	80 μl	160μl	160 μl
20 mM dCTP	80 μl	20 μl	160 μl	160 μl
10 mM 7-dGTP	240 μl	240 μl	120 μl	480 μl
20 mM dTTP	80 μl	80 μl	160 μl	40 μl
5 mM ddATP	1920 μl	–	–	–
5 mM ddCTP	–	960 μl	–	–
5 mM ddGTP	–	–	320μl	–
5 mM ddTTP	–	–	–	3200 ul
50:1 TE	640 ul	640 μl	1280 μl	1280 μl
ddH2O	3420 μl	4380μl	10600 μl	7480 μl
Total	**6400 μl**	**6400 μl**	**12800 μl**	**12800 μl**

To each of these mix solutions, and equal volume of 5X Taq reaction buffer is added (with DMSO), so 6.4 ml is added to A and C, and 12.8 ml is added to G and T. This mix/buffer solution is distributed into 0.5 ml colored microcentrifuge tubes (blue for A, green for C, purple for G, and yellow for T) in 60 or 120 μl aliquots (60 for A and C/120 for G and T). The simplest way to distribute the 60 μl aliquots is 2 x 30 μl using the Eppendorf repeat pipettor set on 3 with the 0.5 ml tips, and for the 120 μl aliquots use 1 x 100 μl with the 5 ml tip plus 1 x 20 μl with the 0.5 ml Combitip. The mixes should be stored at -20 or -70°C.

Assay buffers

10x Low salt buffer
100mM Tris-HCl, pH 7.6
100mM $MgCl_2$
10mM DTT

10x Core buffer
500mM NaCl
500mM Tris-HCl, pH 7.6
100mM $MgCl_2$

10x Medium salt buffer	10x Hind buffer
500mM NaCl	600mM NaCl
100mM Tris-HCl, pH 7.6	100mM Tris-HCl, pH 7.6
100mM $MgCl_2$	70mM $MgCl_2$
10mM DTT	

10x High salt buffer	10x Sma I buffer (1)
1.0M NaCl	200mM KCl
500mM Tris-HCl, pH 7.6	100mM Tris-HCl, pH 7.6
100mM $MgCl_2$	100mM $MgCl_2$
10mM DTT	10mM DTT

APPENDIX-2 : Primers

Forward primer sequence

20mer 5' GACGTTGTAAAACGACGGCC 3'

18mer 5' TGTAAAACGACGGCCAGT 3'

Forward Aminolink-primer sequence

5' -TG TAA AAC GAC GGC CAG T 3'

Reverse primer sequence

20mer 5′ CACAGGAAACAGCTATGACC 3'

18mer 5′ CAGGAAACAGCTATGACC 3'

Reverse Aminolink-primer sequence

5' -CA GGA AAC AGC TAT GAC C- 3'

Oligonucleotide universal primers used for DNA sequencing

The following primers are widely used these days:

- Universal Forward 20mer 5' GTTGTAAAACGACGGCCAGT 3'
- Universal Reverse 20mer 5' CACAGGAAACAGCTATGACC 3'

Reverse primer sequence

- 20mer 5' CACAGGAAACAGCTATGACC 3'
- 18mer 5' CAGGAAACAGCTATGACC 3'
- T7: 5'-TAA-TAC-GAC-TCA-CTA-TAG-GG-3'

- SP6:5'-ATT-TAG-GTG-ACA-CTA-TAG-AA-3'
- M13 (-21) universal forward 5'-TGT-AAA-ACG-ACG-GCC-AGT-3'
- M13 (-40) universal forward 5'-GTT-TTC-CCA-GTC-ACG-AC-3'
- M13/pUC reverse primer 5'-CAG-GAA-ACA-GCT-ATG-ACC-3'
- T7 primer 5'-TAA-TAC-GAC-TCA-CTA-TAG-GG-3'
- SP6 primer 5'-ATT-TAG-GTG-ACA-CTA-TAG-3'
- -16bs 5'-TCG-AGG-TCG-ACG-GTA-TCG-3'
- +19bs 5'-GCC-GCT-CTA-GAA-CTA-GTG-3'

APPENDIX-3 : Units and formulas

Units

- 1 mg = 10-3 g.
- 1 µg = 10-6 g.
- 1 ng = 10-9 g.
- 1 pg = 10-12 g.

1 kb of DNA

- = 6.5 x10+5 Daltons of duplex DNA (sodium salt)
- = 3.3 x10+5 Daltons of single-stranded DNA (sodium salt)
- = 3.4 x10+5 Daltons of single-stranded RNA (sodium salt)

Average MW of a deoxynucleotide base = 324.5 Daltons

Average MW of a deoxynucleotide base pair = 649 Daltons

- 1 µg/ml of DNA = 3.08 uM phosphate
- 1 µg/ml of 1 kb of DNA = 3.08 nM 5' ends
- 1 mol of pBR322 = 2.83 x10+6 g.
- 1 pmol of linear pBR322, 5' ends = 1.4 µg
- 1 A260 unit of duplex DNA = 50 ug
- 1 A260 unit of single-stranded DNA = 37 µg
- 1 A260 unit of single-stranded RNA = 50 µg

1 kb of DNA

- = 333 amino acids of coding capacity
- = 37,000 daltons

Densities (50% GC)

- RF I (supercoiled) ds DNA 1.709 g/ml
- RF II (nicked) ds DNA 1.54 g/ml
- ss DNA 1.726 g/ml
- ss RNA 1.90 g/ml
- protein 1.33 g/ml

APPENDIX-4 : DNA Related Information

DNA melting point

- *For duplex DNA >50 bp*

  Tm = 81.5deg. C +16.6 log (M of NaCl) + 0.41(% GC)

  - [500/bp of shortest chain in duplex]
  - [0.65(% formamide)]

- *For duplex DNA <50 bp*
  - Add 2deg. C for each A or T
  - Add 4deg. C for each G or C

Picomoles of ends

- pmol ends per ug linear DNA = 3030/number of bases

DNA mobility in gels

- Migration of marker dyes in native polyacrylamide non-denaturing gels

Gel %	Bromophenol blue (BP)	Xylene cyanole (XC)
3.5	100	460
5.0	65	260
8.0	45	160
12.0	20	70
20.0	12	45

❒ Migration of marker dyes in polyacrylamide denaturing gels

Gel %	Bromophenol blue (BP)	Xylene cyanole (XC)
5.0	35	130
6.0	26	106
8.0	19	75
10.0	12	55
20.0	8	28

❒ Relative positions of different DNA forms on Tris-acetate agarose gels

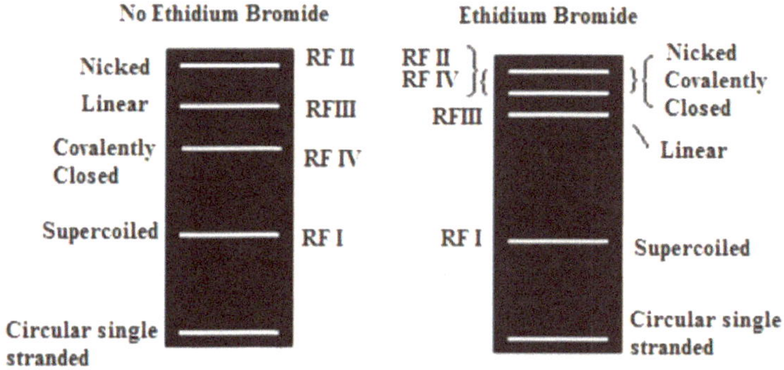

The exact distance between bands is influenced by percentage of agarose, time of electrophoresis, concentration of Ethidium bromide, degree of supercoiling and the size and complexity of the DNA.

Codon chart and amino acid symbols Second base

First Base	U	C	A	G	Third Base
U	UUU Phe	UCU Ser	UAU Tyr	UGU Cys	U
	UUC	UCC	UAC	UGC	C
	UUA Leu	UCA	UAA Ocr	UGA Opl	A
	UUG	UCG	UAG Amb	UGG Trp	G
C	CUU Leu	CCU Pro	CAU His	CGU Arg	U
	CUC	CCC	CAC	CGC	C
	CUA	CCA	CAA Gln	CGA	A
	CUG	CCG	CAG	CGG	G
A	AUU Ile	ACU Thr	AAU Asn	AGU Ser	U
	AUC	ACC	AAC	AGC	C
	AUA	ACA	AAA Lys	AGA Arg	A
	AUG Met	ACG	AAG	AGG	G
G	GUU Val	GCU Ala	GAU Asp	GGU Gly	U
	GUC	GCC	GAC	GGC	C
	GUA	GCA	GAA Glu	GGA	A
	GUG	GCG	GAG	GGG	G

APPENDIX-5 : Amino Acids

One and three-letter amino acid symbols

Alanine	**Ala**	**A**
Arginine	Arg	R
Asparagine	Asn	N
Aspartic acid	Asp	D
Asparagine or Aspartic acid	Asx	B
Cysteine	Cys	C
Glutamine	Gln	Q
Glutamic acid	Glu	E
Glutamine or Glutamic acid	Glx	Z
Glycine	Gly	G
Histidine	His	H
Isoleucine	Ile	I
Leucine	Leu	L
Lysine	Lys	K
Methionine	Met	M
Phenylalanine	Phe	F
Proline	Pro	P
Serine	Ser	S
Threonine	Thr	T
Tryptophan	Trp	W
Tyrosine	Tyr	Y
Valine	Val	V

Glossary

Agar or agar agar

It is a gelatinous substance derived from red algae, belonging to genera *Gelidium* and *Gracilaria,* or seaweed (*Sphaerococcus euchema*). It has found extensive use as a solid substrate to contain culture medium for microbiological work. The gelling agent is an unbranched polysaccharide obtained from the cell walls of the red algae. Chemically, agar is a polymer made up of subunits of the sugar galactose. Agar polysaccharides serve as the primary structural support for the algae's cell walls.

Agrobacterium tumefaciens

It is the causal agent of crown gall disease (the formation of tumours) in over 140 species of dicot. It is a rod shaped,Gram negative soil bacterium. Symptoms are caused by the insertion of a small segment of DNA (known as the T-DNA, for 'transfer DNA') into the plant cell, which is incorporated at a semi-random location into the plant genome.

Agrobacterium tumefaciens (or *A. tumefaciens*) is a member of the family Rhizobiaceae, which includes the nitrogen fixing legume-symbionts. *Agrobacterium* are pathogenic and do not benefit the plant. Economically, *A. tumefaciens* is a serious pathogen of walnuts, grape vines, stone fruits, nut trees, sugar beets, horse radish and rhubarb.

Alkaline phosphatase (ALP)

It is a hydrolase enzyme responsible for removing phosphate groups from many types of molecules, including nucleotides, proteins, and alkaloids. The process of removing the phosphate group is called *dephosphorylation*. As the name suggests, alkaline phosphatases are most effective in an alkaline environment.

Bacterial Transformation and Transfection

It is the process by which bacterial cells take up naked DNA molecules. If the foreign DNA has an origin of replication recognized by the host cell DNA polymerases, the bacteria will replicate the foreign DNA along with their own DNA. When transformation is coupled with antibiotic selection techniques, bacteria can be induced to uptake certain DNA molecules, and those bacteria can be selected for that incorporation. Bacteria which are able to uptake DNA are called "competent" and are made so by treatment with calcium chloride in the early phase of growth. The bacterial cell membrane is permeable to chloride ions, but is non-permeable to calcium ions. As the chloride ions enter the cell, water molecules accompany the charged particle. This influx of water causes the cells to swell and is necessary for the uptake of DNA. The exact mechanism of this uptake is unknown. It is known, however, that the calcium chloride treatment be followed by heat. When *E. coli* are subjected to 42°C heat, a set of genes are expressed which aid the bacteria in surviving at such temperatures. This set of genes is called the heat shock genes. The heat shock step is necessary for the uptake of DNA. At temperatures above 42°C, the bacteria's ability to uptake DNA becomes reduced, and at extreme temperatures the bacteria will die.

Bacteriophage M13 Transformation and Viral Transfection

The transformation of bacteriophage M13 into bacterial cells is identical to plasmid DNA transformation through the heat shock step. After the heat shock step, single stranded M13 DNA begins replicating in the host cell through use of the host cell machinery. During the life cycle of this virus, however, M13 replicative form is created and daughter phages are packaged and extruded from the bacterial cell. These intact phage molecules then infect neighboring bacteria in a process called transfection. When these transformed and transfected bacteria are plated with non-infected cells onto growth media, the non-infected cells form a background cell lawn which covers the plate. In regions of M13 transfection, areas of slowed growth, called plaques, can be identified as opaque regions which interrupt the lawn.

Bacterial Strains

Since M13 viral transfection is a critical part of the transformation of bacterial cells with M13, it is absolutely necessary to use a strain of E. coli which harbors the episome for the F pilus. When M13 phages infect bacterial cells they attach to the F pilus, and the loss of this pilus is a common reason for a failed or poor transformation/transfection of M13. JM101 is a strain of E. coli which possesses the F pilus if the culture is maintained under appropriate conditions. Since the F pilus is not necessary for plasmid DNA transformation, it is advisable to use GM272, a much healthier, F- strain of E. coli for this procedure. To avoid confusion between the similar procedures, bacterial transformation with plasmid DNA is termed a "Transformation", and a bacterial transformation with naked M13 followed by a transfection with intact M13 phage is called a "Transfection."

BCIP & NBT

Produce blue-purple precipitates for Western blotting or staining applications

An ideal system for blotting or staining applications is the combination of BCIP (5-Bromo-4-Chloro-3'-Indolyphosphate p-Toluidine Salt) and NBT (Nitro-Blue Tetrazolium Chloride). Together they yield an intense, insoluble black-purple precipitate when reacted with Alkaline Phosphatase. The NBT/BCIP reaction is illustrated in the figure below. This reaction proceeds at a steady rate, allowing accurate control of the relative sensitivity and control of the development of the reaction.

NBT/BCIP reaction

BCIP is hydrolyzed by alkaline phosphatase to form an intermediate that undergoes dimerization to produce an indigo dye. The NBT is reduced to the NBT-formazan by the two reducing equivalents generated by the dimerization.

- Ideal for immunoblotting
- Ready-to-use single component
- Minimal assay-to-assay variability
- Sensitive with low background

BCIP/NBT plus Suppressor is a one-component solution, which can be used for staining tissues assayed with calf intestinal alkaline phosphatase without worrying about the interference caused by endogenous phosphatases. This sensitive substrate formulation contains 1 mM levamisole to inhibit endogenous phosphatase activity, while yielding an intense dark purple precipitate in the presence of calf intestinal phosphatase. Low backgrounds result due to the suppression of endogenous phosphatase activity.

Dithiothreitol (DTT)

It is the common name for a small-molecule redox reagent known as Cleland's reagent. DTT's formula is $C_4H_{10}O_2S_2$. A common use of DTT is as a reducing or "deprotecting" agent for thiolated DNA. The terminal sulfur atoms of thiolated DNA have a tendency to form dimers in solution, especially in the presence of oxygen. Typically DTT is mixed with a DNA solution and allowed to react, and then is removed by filtration (for the solid catalyst) or by chromatography (for the liquid form). The DTT removal procedure is often called "desalting."

DTT is frequently used to reduce the disulfide bonds of proteins and, more generally, to prevent *intramolecular* and *intermolecular* disulfide bonds from forming between cysteine residues of proteins.

DMSO

Dimethyl sulfoxide (DMSO) is the organosulfur compound with the formula $(CH_3)_2SO$. This colorless liquid is an important polar solvent that dissolves both polar and nonpolar compounds and is miscible in a wide range of organic solvents as well as water. It has a distinctive property of penetrating the skin very readily, so that one may taste it soon after it comes into contact with the skin. Its taste has been described as oyster- or garlic-like.

DMSO is an important polar solvent. Because of its excellent solvating power, DMSO is frequently used as a solvent for chemical reactions involving salts. It is also extensively used as an extractant in biochemistry and cell biology. Because DMSO is only weakly acidic, it tolerates relatively strong bases and as such has been extensively used in the study of carbanions. It has a high boiling point, hence DMSO evaporates slowly at normal atmospheric pressures. Reactions conducted in DMSO are often diluted with water to precipitate or phase-separate products. DMSO is an effective paint stripper. The relatively high freezing point of DMSO hence at room temperature it is a solid.

In its deuterated form (DMSO-d_6), it is a useful but expensive solvent for NMR spectroscopy, again due to its ability to dissolve a wide range of analytes, its own simple spectrum, and its suitability for high-temperature NMR spectroscopic studies. DMSO is finding increased use in manufacturing processes to produce microelectronic devices. It is widely used to strip photoresist in TFT-LCD 'flat panel' displays and advanced packaging applications (such as wafer-level packaging / solder bump patterning). It also used in biopreservation especially stem cell banking.

DMSO is used in PCR to inhibit secondary structures in the DNA template or the DNA primers. It is added to the PCR mix before reacting, where it interferes with the self-complementarity of the DNA, minimizing interfering reactions. However, use of DMSO in PCR increases the mutation rate. DMSO may also be used as a cryoprotectant, added to cell media to prevent cell death during the freezing process.

EDTA

Ethylenediaminetetraacetic acid is a widely used organic compound . The conjugate base is named ethylenediaminetetraacetate. EDTA is a polyamino carboxylic acid and a colourless, water-soluble solid. It is widely used to dissolve scale. Its usefulness arises because of its role as a chelating agent, i.e. its ability to "sequester" metal ions such as Ca^{2+} and Fe^{3+}. After being bound by EDTA, metal ions remain in solution but exhibit diminished reactivity. EDTA is produced as several salts, notably disodium EDTA and calcium disodium EDTA.

In the laboratory, EDTA is widely used for scavenging metal ions: In biochemistry and molecular biology, ion depletion is commonly used to deactivate metal-dependent enzymes, either as an assay for their reactivity or to suppress damage to DNA or proteins. In analytical chemistry, EDTA is used in complexometric titrations and analysis of water hardness or as a masking agent to sequester metal ions that would interfere with the analyses. EDTA finds many specialized uses in the biomedical laboratories, such as in veterinary ophthalmology as ananticollagenase to prevent the worsening of corneal ulcers in animals. In tissue culture EDTA is used as a chelating agent that binds to calcium and prevents joining of cadherins between cells, preventing clumping of cells grown in liquid suspension, or detaching adherent cells for passaging.In histopathology, EDTA can be used as a decalcifying agent making it possible to cut sections using a microtome once the tissue sample is demineralised.

Klenow fragment

It is a large protein fragment produced when DNA polymerase-I from *E. coli* is enzymatically cleaved by the protease subtilisin. First reported in 1970, it retains the 5'-3' polymerase activity and the 3′ ? 5′ exonuclease activity for removal of precoding nucleotides and proofreading, but loses its 5' ? 3' exonuclease activity.

The other smaller fragment formed when DNA polymerase I from *E. coli* is cleaved by subtilisin retains the 5'-3' exonuclease activity but does not have the other two activities exhibited by the Klenow fragment (i.e. 5'-> 3' polymerase activity, and 3'->5' exonuclease activity).

LB Medium

Lysogeny broth (LB), a nutritionally rich medium, is primarily used for the growth of bacteria. It is also known as Luria broth or Luria-Bertani broth, though both of these names are incorrect. The original medium was developed by Bertani alone and was named "lysogeny broth" in his original paper.

LB media formulations have been an industry standard for the cultivation of *Escherichia coli* as far back as the 1950s. These media have been widely used in molecular microbiology applications for the preparation of plasmid DNA and recombinant proteins. It continues to be one of the most common media used for maintaining and cultivating recombinant strains of *Escherichia coli*.

There are several common formulations of LB. Although they are different, they generally share a somewhat similar composition of ingredients used to promote growth, including the following:

- Peptides and casein peptones
- Vitamins (including B vitamins)
- Trace elements (e.g. nitrogen, sulfur, magnesium)
- Minerals

Peptides and peptones are provided by tryptone. Vitamins and certain trace elements are provided by yeast extract. Sodium ions for transport and osmotic balance are provided by sodium chloride. Bacto-tryptone is used to provide essential amino acids to the growing bacteria, while the bacto-yeast extract is used to provide a plethora of organic compounds helpful for bacterial growth.

LB Medium	
Bacto-Tryptone	10g
Bacto-yeast extract	5g
NaCl	10g
ddH_2O	1litre
Total Vol.	1 litre

Adjust pH to 7.0 and autoclave to sterilize

MOPS

MOPS are the common name for the compound 3-(N-morpholino) propanesulfonic acid a buffer introduced by Good et al. in the 1960s. Its chemical structure contains a morpholine ring. With a pK_a of 7.20, MOPS is an excellent buffer for many biological systems at near-neutral pH.

MOPS is frequently used as a buffering agent in biology and biochemistry. It has been tested and recommended for polyacrylamide gel electrophoresis. Usage above 20 mM in mammalian cell culture work is not recommended.

Proteinase K

Proteinase K is a subtilisin-like endolytic protease that is isolated from the saprophytic fungus*Tritirachium album*. It has a high activity that is stable across a wide range of pH and temperature conditions and is suited to short digestion times. The activity of proteinase K is increased at elevated temperatures up to 65°C. Calcium is not essential to the function of proteinase K. Therefore, EDTA and other chelating agents do not interfere with the activity and may be used alongside proteinase K to inactivate calcium-dependent nucleases in DNA and RNA preparation.

RNAse A

Ribonuclease A (RNase A) is an endoribonuclease that cleaves single-stranded RNA at the 3′ end of pyrimidine residues, forming oligoribonucleotides having 3′-terminal pyrimidine-3′-phosphates. Pyrimidine-3′-monophosphates are also released by RNAse A cleavage of adjacent pyrimidine nucleotides.

Plasmid Transformation and Antibiotic Selection

The process for the uptake of naked plasmid and bacteriophage DNA is the same; calcium chloride treatment of bacterial cells produces competent cells which will uptake DNA after a heat shock step. However, there is a slight, but important difference in the procedures for transformation of plasmid DNA and bacteriophage M13 DNA. In the plasmid transformation, after the heat shock step intact plasmid DNA molecules replicate in bacterial host cells. To help the bacterial cells recover from the heat shock, the cells are briefly incubated with non-selective growth media. As the cells recover, plasmid genes are expressed, including those that enable the production of daughter plasmids which will segregate with dividing bacterial cells. However, due to the low number of bacterial cells which contain the plasmid and the potential for the plasmid not to propogate itself in all daughter cells, it is necessary to select for bacterial cells which contain the plasmid. This is commonly performed with antibiotic selection. E. coli strains such as GM272 are sensitive to common antibiotics such as ampicillin. Plasmids used for the cloning and manipulation of DNA have been engineered to harbor the genes for antibiotic resistance. Thus, if the bacterial transformation is plated onto media containing ampicillin, only bacteria which possess the plasmid DNA will have the ability to metabolize ampicillin and form colonies. In this way, bacterial cells containing plasmid DNA are selected.

Plasmid Transformation and Antibiotic Selection Lac Z Operon

An additional level of selection can be achieved during transformation and transfections. Bacterial cells containing plasmids with the antibiotic resistance gene are selected in bacterial transformations, and cells in an area of M13 infection are recognized as plaques against a lawn of non-infected cells. However, the object of most transformations and transfections is to clone foreign DNA of interest into a known plasmid or viral vector and to isolate cells containing those recombinant molecules from each other and from those containing the non-recombinant vector. The E. coli lacZ operon has been incorporated into several cloning vectors, including plasmid pUC and bacteriophage M13. The polylinker regions of these vectors were engineered inside of the lacZ gene coding region, but in a way not to interrupt the reading frame or the functionality of the resultant lacZ gene protein product. This protein product is a galactosidase. In recombinant vectors which have an insert DNA molecule cloned into one of the restriction enzyme sites in the polylinker, this insert DNA results in an altered lacZ gene and a non-functional galactosidase. The presence or absence of this protein can easily be determined through the use of a simple chromogenic assay using IPTG and X-Gal. IPTG is the lacZ gene inducer and is necessary for the production of the galactosidase. The usual substrate for the lacZ gene protein product is galactose, which is metabolized into lactose and glucose. X-Gal is a colorless, modified galactose sugar. When this molecule is metabolized by the galactosidase, the resultant products are a bright blue color.

When IPTG and X-Gal are included in plasmid DNA transformation, blue colonies represent bacteria harboring non-recombinant pUC vector DNA since the lacZ gene region is intact. IPTG induces production of the functional galactosidase which cleaves X-Gal and results in a blue colored metobolite. It follows that colorless colonies contain recombinant pUC DNA since a nonfunctional galactosidase is induced by IPTG which is unable to cleave the X-Gal. Similarly, for bacteriophage transfections, colorless plaques indicate regions of infection with recombinant M13 viruses, and blue plaques represent infection with non-recombinant M13.

TAE Buffer

It is a buffer solution containing a mixture of Tris base, acetic acid and EDTA. In molecular biology it is used in agarose electrophoresis typically for the separation of nucleic acids such as DNA and RNA.[1] It is made up of Tris-acetate buffer, usually at pH 8.0, and EDTA, which sequesters divalent cations. TAE has a lower buffer capacity than TBE, but linear, double stranded DNA runs faster in TAE.

TAE (Tris-acetate-EDTA) buffer is used as both a running buffer and in agarose gel. Its use in denaturing gradient gel electrophoresis methods for broad-range mutation analysis has also been described.

TBE or Tris/Borate/EDTA

It is a buffer solution containing a mixture of Tris base, boric acid and EDTA. In molecular biology, TBE and TAE buffers are often used in procedures involving nucleic acids, the most common being electrophoresis. Tris-acid solutions are effective buffers for slightly basic conditions, which keep DNA deprotonated and soluble in water. EDTA is a chelator of divalent cations, particularly of magnesium (Mg^{2+}). As these ions are necessary co-factors for many enzymes, including contaminant nucleases, the role of the EDTA is to protect the nucleic acids against enzymatic degradation. But since Mg^{2+} is also a co-factor for many useful DNA-modifying enzymes such as restriction enzymes and DNA polymerases, its concentration in TBE or TAE buffers is generally kept low (typically at around 1 mM).

Preparation of 1 liter of 5X stock solution

- 53 g of Tris base
- 27.5 g of
- 20 ml of 0.5 M EDTA (pH 8.0)

TBE can be diluted to 0.5X prior to use in electrophoresis, 1x is acceptable as well.

Tetramethylethylenediamine (TMEDA or TEMED)

It is a chemical compound with the formula $(CH_3)_2NCH_2CH_2N(CH_3)_2$. This species is derived from ethylenediamine by replacement of the four N-H groups with four N-methyl groups. It has a disagreeable fishy odor.

TMEDA is widely employed as a ligand for metal ions. It forms stable complexes with many metal halides, e.g. zinc chloride and copper (I) iodide, giving complexes that are soluble in organic solvents. In such complexes, TMEDA serves as a bidentate ligand.

Tetramethylethylenediamine is used with ammonium persulfate to catalyze the polymerization of acrylamide when making polyacrylamide gels, used in gel electrophoresis, for the separation of proteins or nucleic acids. Although the amounts used in this technique may vary from method to method, 0.1-0.2% v/v TMEDA is a "traditional" range.

Ti plasmid

It is a circular plasmid that often, but not always, is a part of the genetic equipment that *Agrobacterium tumefaciens* and *Agrobacterium rhizogenes* use to transduce its genetic material to plants. The Ti plasmid is lost when Agrobacterium is grown above 28C. Such cured bacteria do not induce crown galls, i.e. they become virulent. The Ti plasmids are classified into different types based on the type of opine produced by their genes.

The plasmid has 196 genes that code for 195 proteins. The plasmid is 206,479 nucleotides long, the GC content is 56% and 81 per cent of the material are coding genes. There are no pseudo genes. The modification of this plasmid is very important in the creation of transgenic plants, but only in dicotyledon plants.

Tris

Tris (also known as THAM) is an abbreviation of the organic compound known as tris (hydroxymethyl) aminomethane, with the formula $(HOCH_2)_3CNH_2$. Tris is extensively used in biochemistry and molecular biology. Tris is widely used as a component of buffer solutions, such as in TAE and TBE buffer, especially for solutions of nucleic acids. It is a primary amine and thus undergoes the reactions associated with typical amines, e.g. condensations with aldehydes.

Tris Buffered Saline (TBS)

It is a buffer used in some biochemical techniques to maintain the pH within a relatively narrow range. Tris (with HCl) has a slightly alkaline buffering capacity in the 7 - 9.2 range. Tris has a pK_a of 8.06 at 20 °C. The pK_a declines approximately 0.03 units per degree Celsius rise in temperature. Thus, shifting the temperature can lead to pH-alteration.

Preparation of TBS

- 50 mM Tris
- 150 mM NaCl

Adjust pH with HCl to pH 7.6.

The simplest way to prepare a TBS solution is to use TBS tablets. They are formulated to give a ready to use TBS solution upon dissolution in 500 ml of deionized water.

TBS-Tween tablets are also available.

TSS Buffer

Transformation & Storage Solution (TSS), developed at the U.S. National Institute of Health, enables researchers to prepare competent *E. coli* in a single step and to transform the cells without heat-shock. TSS has been reported to be faster and easier than other methods of producing competent cells, such as the traditional $CaCl_2$ method described by Sam brook *et al.* or other high-competency protocols.

Transformation efficiencies depend on the strain of *E. coli*, as well as the nature and quality of the transforming DNA.

To make 50 mL:

- 5g PEG 8000
- 1.5 mL 1M $MgCl_2$ (or 0.30g $MgCl_2$*$6H_20$)
- 2.5 mL DMSO
- Add LB to 50 mL

Filter sterilize

Protocol for using TSS for transforming *E. coli*

1. Pellet 1 ml of cells at A600 ~ 0.3 in 1.5 ml microcentrifuge tubes.
2. Remove the supernatant, add 0.1 ml of ice-cold 1X TSS, and gently resuspend the cells. Cells can be frozen at this point and stored at -70°C for later use.
3. Add 100 pg-10 ng of transforming DNA.
4. Mix the cells and DNA and place on ice for 10 minutes, then at room temperature for 10 minutes and then again on ice for 10 minutes.
5. Add 1 ml of LB broth and incubate the tubes at 37°C for 1 hour.
6. Plate the cells on the appropriate selective or differential media.

T4 DNA Polymerase

T4 DNA Polymerase catalyzes the synthesis of DNA in the 5′→ 3′ direction and requires the presence of template and primer. This enzyme has a 3′→ 5′ exonuclease activity which is much more active than that found in DNA Polymerase I. Unlike *E. coli* DNA Polymerase I, T4 DNA Polymerase does not have a 5′→ 3′ exonuclease function. It can be purified from a strain of *E. coli* that carries a T4 DNA Polymerase overproducing plasmid.

Applications

- 3′ overhang removal to form blunt ends.
- 5′overhang fill-in to form blunt ends.
- Single strand deletion subcloning.
- Second strand synthesis in site-directed mutagenesis.
- Probe labeling using replacement synthesis

T4 DNA Ligase

Catalyzes the formation of a phosphodiester bond between juxtaposed 5' phosphate and 3' hydroxyl termini in duplex DNA or RNA. This enzyme will join blunt end and cohesive end termini as well as repair single stranded nicks in duplex DNA, RNA or DNA/RNA hybrids. It can be purified from *E. coli*. It is used in cloning of restriction fragments and joining linkers and adapters to blunt-ended DNA.

References

- ABI 392 Synthesizer manual.
- Adams, S.P., Kavka, K.S., Wykes, E.J., Holder, S.B., and Galluppi, G.R., "Hindered dialkylamino nucleoside phosphite reagents in the synthesis of two DNA 51-mers", J. Am. Chem. Soc. 105, 661 (1983).
- Bachman, B.J. (1990) Microbiology Reviews 54: 130- 197.
- Bankier, A.T., Weston, K.M., and Barrell, B.G., "Random cloning and sequencing by the M13/dideoxynucleotide chain termination method", Meth. Enzymol. 155, 51-93 (1987).
- Beaucage, S.L. and Caruthers, M., "Deoxynucleoside phosphoramidites: A new class of key intermediates for deoxypolynucleotide synthesis", Tetrahedron Letters 22, 1859-1862 (1981).
- Beckman Oligo 1000 Synthesizer manual.
- Bethesda Research Laboratories, 1994 catalog.
- Birnboim, H.C. and Doly, J., "A rapid alkaline extraction procedure for screening recombinant plasmid DNA", Nucl. Acids Res. 7, 1513-1523 (1979).

- Bruce A. Roe, Department of Chemistry and Biochemistry, The University of Oklahoma, Norman, Oklahoma 73019 broe@ou.edu
- Carter, J.M. and Milton, I.D., "An inexpensive and simple method for DNA purifications on silica particles", Nucl. Acids Res. 21, 1044 (1993).
- Cearley JA, Bolyard MG: Regeneration of *Solanum tuberosum* cv. Katahdin from leaf explants *in vitro*. Am Potato J 74: 125-129 (1997).
- Chissoe, S.L., Wang, Y.F., Clifton, S.W., Ma, N., Sun, H.J., Lobsinger, J.S., Kenton, S.M., White, J.D., and Roe, B.A., "Strategies for rapid and accurate DNA sequencing", Methods: Companion Methods Enzymol. 3, 55-65 (1991).
- Cohen, S.N., Chang, A.C.Y., and Hsu, L, "Nonchromosomal antibiotic resistance in bacteria: genetic transformation of *Escherichia coli* by R-factor DNA", Proc. Natl. Acad. Sci. U.S.A. 69, 2110-2114 (1972).
- Davis, R.W., D. Botstein and J.R. Roth, A Manual for Genetic Engineering: Advanced Bacterial Genetics. Cold Spring Harbor Laboratories, Cold Spring Harbor, N.Y. 1980.
- Eadie, J.S. and Davidson, D.S., "Guanine modification during chemical DNA synthesis", Nucl. Acids Res. 15, 8333-8349 (1987).
- Engler, M.J. and Richardson, C.C. (1982) In: The Enzymes, Boyer, P.D., ed., AcademicPress, New York, NY.4. Protocols and Applications Guide, Third Edition (1996) Promega Corporation.
- Eperon, I.C.,"Rapid preparation of bacteriophage DNA for sequence analysis in sets of 96 clones, using filtration", Anal. Biochem. 156, 406-412 (1986).
- Farrance, I.K., Eadie, J.S., and Ivarie, R.,"Improved chemistry for oligodeoxyribonucleotide synthesis substantially improves restriction enzyme cleavage of a synthetic 35mer", Nucl. Acids Res. 17, 1231-1245 (1989).
- Federal Bureau of Investigation, RFLP Manual, U.S. Government 1993.
- GS Gene Prep Manifold Instruction Manual, BioRad Labs, Richmond, CA
- Hanahan (1983) J. Mol. Biol. 166, 557-580.
- Huynh, Young, and Davis (1985) DNA Cloning, *Vol. 1*, 56-110.
- Ish-Horowicz, D. and Burke, J.F., "Rapid and efficient cosmid cloning", Nucl. Acids Res. 9, 2989 (1981).
- Jerpseth, J., et al.,"XL1-Blue MRF' E. coli cells: McrA-, McrCB-, McrF-, Mrr-, HsdR- derivative of XL1-Blue cells" Strategies 6, 81 (1992).

- Johnston-Dow, L., Mardis, E., Heiner, C., and Roe, B.A., "Optimized methods for fluorescent and radiolabeled DNA sequencing", BioTechniques 5, 754-765 (1987).
- Kusukawa, N., Uemori, T., Asada, K., and Kato, I., "Rapid and reliable protocol for direct sequencing of material amplified by the polymerase chain reaction", BioTechniques 9, 66-72 (1990).
- Lee, L.G., Connell, C.R., Woo, S.L., Cheng, R.D., McArdle, B.F., Fuller, C.W., Halloran, N.D., and Wilson, R.K., "DNA sequencing with dye-labeled terminators and T7 DNA polymerase: effects of dyes and dNTPs on incorporation of dye-terminators and probability analysis of termination fragments", Nucl. Acids Res. 20, 2471-2483 (1992).
- Mandel, M. and Higa, A., "Calcium dependent bacteriophage DNA infection", J. Mol. Biol. 53, 154-159 (1970).
- Mardis, E.R. and Roe, B.A., "Automated methods for single-stranded DNA isolation and dideoxynucleotide DNA sequencing reactions on a robotic workstation", BioTechniques 7, 840-850 (1989).
- Matteucci, M.D. and Caruthers, M.H., "Synthesis of deoxynucleotides on a polymer support". J. Am. Chem. Soc. 103, 3185 (1981).
- Messing, J., "New M13 vectors for cloning", Meth. Enzymol. 101, 20-78 (1983).
- Mullis, K.B. and Faloona, F.A., "Specific synthesis of DNA *in vitro* via a polymerase catalyzed chain reaction", Meth. Enzymol. 155, 335-350 (1987).
- Murray, N.E.,Brammar, W.J., and Murray, K., "Lamphoid phages that simplify the recovery of in vitro recombinants", Mol. Gen. Genet. 150, 53-61 (1977).
- Pan, H., Chissoe, S. L., Bodenteich, A., Wang, Z., Iyer, K., Clifton, S.W., Crabtree, J.S., and Roe, B. A. "The complete nucleotide sequences of the SacBII Kan domain of the P1 and pAD10-SacBII cloning vector and three cosmid vectors: pTCF, svPHEP, and LAWRIST16", GATA 11 (5-6): 181-186 (1994).
- Pheiffer, B.H. and Zimmerman, S.B. (1983) Polymer stimulated ligation: Enhanced blunt- or cohesive-end ligation of DNA or deoxyribooligonucleotides by T4 DNA ligasein polymer solutions. Nucl. Acids Res. 1 11 1, 7853–71.
- Protocol included with the ABI PRISM Sequenase[TM] Terminator Double-Stranded DNA Sequencing Kit, part number 401461.
- Protocol included with the ABI PRISM Sequenase[TM] Terminator Single-Stranded DNA Sequencing Kit, part number 401458.

- Richardson, C.C., "Polynucleotide kinase from *Escherichia coli* infected with bacteriophage T4" in Procedures in nucleic acids research (ed. G.L. Cantoni and D.R. Davies), vol. 2, p. 815, Harper and Row, New York (1971).
- Sambrook, J. and D. Russell. 2001. "Molecular Cloning: A Laboratory Manual", 3rd edition. Cold Spring Harbor Laboratory Press. Cold Spring Harbor, NY. p. 7.32.
- Sambrook, J., Fritsch, E.F. and Maniatis, T. (1989) Molecular Cloning: A LaboratoryManual, Second Edition, Cold Spring Harbor Laboratory Press, Cold Spring Harbor,NY.
- Sambrook, J., Fritsch, E.F., and Maniatis, T., in Molecular Cloning: A Laboratory Manual. Cold Spring Harbor Laboratory Press, NY, Vol. 1,2,3 (1989).
- Sanger, F., Coulson, A.R., Barrell, B.G., Smith, A.J.H., and Roe, B.A., "Cloning in Single-stranded Bacteriophage as an Aid to Rapid DNA Sequencing", J. Mol. Biol. 143, 161-178 (1980).
- Sawadogo, M. and Van Dyke, M.W., "A rapid method for the purification of deprotected oligodeoxynucleotides", Nucl. Acids Res. 19, 674 (1991).
- Siemenak, D.R., Sieu, L.C., and Slightom, J.L., "Strategy and methods for directly sequencing cosmid clones", Anal. Biochem 192, 441-448 (1991).
- Smith, L.M., Kaiser, R.J., Sanders, J.Z., and Hood, L.E., "The synthesis and use of fluorescent oligonucleotides in DNA sequence analysis", Meth. Enzymol. 155, 260-301 (1987).
- Starr, L., and Quaranta, V., "An efficient and reliable method for cloning PCR-amplification products: A survey of point mutations in integrin cDNA", BioTechniques 13, 612-618 (1992).
- Studier, F.W., "Analysis of bacteriophage T7 early RNAs and proteins on slab gels", J. Mol. Biol. 79, 237-248 (1973).
- Tahara, T., Kraus, J.P., and Rosenberg, L.E., "Direct DNA sequencing of PCR amplified genomic DNA by the Maxam-Gilbert method", BioTechniques 8, 366-368 (1990).
- Willis, E.H., Mardis, E.R., Jones, W.L., and Little, M.C., "Prep-A-Gene [TM]: A superior matrix for the purification of DNA and DNA fragments", BioTechniques 9, 92-99 (1990).
- Woodcock, D.M., Crowther, D.M., Doherty, J., Jefferson, S., DeCruz, E., Noyer-Weidner, M., Smith, S.S., Michael, M.Z., and Graham, M.W., "Quantitative

evaluation of *Escherichia coli* host strains for tolerance to cytosine methylation in plasmid and phage recombinants", Nucl. Acids Res. 17, 3469-3478 (1989).

- Yanisch-Perron, C., Viera, J., and Messing, J., "Improved M13 phage cloning vectors and host strains: nucleotide sequences of the m13mp18 and pUC19 vectors", Gene 33, 103-199.

Index

FSC
www.fsc.org
MIX
Papier | Fördert gute Waldnutzung
FSC® C083411

Zeitfracht Medien GmbH
Ferdinand-Jühlke-Straße 7
99095 Erfurt, Deutschland
produktsicherheit@kolibri360.de